Defending the World

Defending the World

The Politics and Diplomacy of the Environment

David Adamson

I.B. TAURIS & CO. LTD
Publishers
London · New York

Published in 1990 by
I.B. Tauris & Co. Ltd
110 Gloucester Avenue
London NW1 8JA

175 Fifth Avenue
New York
NY 10010

In the United States of America
and Canada distributed by
St Martin's Press
175 Fifth Avenue
New York
NY 10010

British Library Cataloguing in Publication Data
Adamson, David
  Defending the World.
  1. Environment. Policies of government
  I. Title
  333.7

  ISBN 1-85043-302-X

Library of Congress Catalog Card Number 90-071483
A full CIP record is available from the Library of Congress

Printed in Great Britain by
Redwood Press Limited, Melksham, Wiltshire

Contents

	Introduction	1
1	Setting the agenda	9
2	Ideals and realities	23
3	Climatology, survival and politics	37
4	The Global Commons: the atmosphere	50
5	The Global Commons: the oceans	61
6	The Global Commons: Antarctica	75
7	The Global Commons: the tropical forests	90
8	The ozone layer	105
9	Climate change and agriculture: and a Soviet dissent	117
10	Eco-politics: the greening of the greys	131
11	The campaigners	144
12	The uncertain politics of carbon dioxide	154
13	Nuclear power: carbon-free but costly and feared	170
14	The pioneering Netherlands: cows, cars and the Western world's age of mobility	185
15	Cleaning up Eastern Europe	199
16	1992 and beyond	212
	Explanatory notes	227
	Select bibliography	233
	Index	235

Introduction

I began this book on the premise that the last decade of the twentieth century would see nuclear arms control gradually superseded by climate change as the first item on the world's agenda for action. Premises of that sort rarely remain intact and this one is no exception. I have knocked out 'gradually'. The pace of change has been much faster than I expected. And for three main reasons. The first is an obvious one: at the beginning of the 1990s governments may still have their doubts about the causes of global warming and its consequences, but they have accepted that scientists are not scaremongering and that the emission of greenhouse gases must be curbed. As far as they can be. The other reasons are the ending of the cold war and the greening of the political mainstream in the West; but more of them later.

This, as I say in the Explanatory Notes, is not a scientific book; evaluating scientific opinions on the greenhouse effect is not its business. The reader should be aware, though, that scientists are unlikely to produce firm answers on climate warming in the short term. None of the five computers working for the Intergovernmental Panel on Climate Change is powerful enough to crack the many complexities of climate change, refine them and reassemble them in a pattern that entirely satisfies the demanding minds of the scientists. What is certain is that climate change seems to correlate with industrial activity. And the evidence of harm is strong enough to support the campaign for a global climate convention. Whether such a convention can be effective in limiting the emission of the most important and pervasive greenhouse gas, carbon dioxide, has to be doubted though. A second member of the greenhouse pantheon, methane, is also unlikely to prove susceptible to reduction by treaty.

President Bush's ambition to go down in history as a 'green President' may have suffered a setback as a result of the hedged terms in which he addressed two important meetings in the first half of 1990 (the IPCC meeting in Washington in February, and the international conference on global warming convened in the same city in April); but at least he caught the mood of ambivalence that prevailed at the beginning of the 1990s: 'All of us must make certain that we preserve our environmental well-being and our economic welfare,' he said at the April conference. 'We know that these are not separate concerns, they are two sides of the same coin.' America's goal remained that set out at the IPCC meeting of 'matching policy commitments to emerging scientific knowledge, and a reconciling of environmental protection to the continued benefits of economic development.' One billion dollars – a 60 per cent increase – would be spent on the United States' climate change research programme. In the meantime, while the scientists worked out what was happening to the satisfaction of the White House and industrialists, the United States would continue to combine its position as the world's largest economy with that of the world's biggest emitter of carbon dioxide.

Whether economic (and population) growth can go hand in hand with a clean and healthy world is an issue contained within the term 'sustainable development'. The political triumph of the West has been built on its economic success in the post-war years. All have become rich. They have survived oil shocks and the loss of empires with hardly more than a tremor in their growth charts. Elsewhere, however, are some forty countries whose inhabitants have become poorer during the past two or three decades. And poverty, or at least not getting richer, could spread.

The second reason why the speed of change is so fast concerns a different kind of global warming, the result of the remarkable – and one hopes, permanent – changes that have occurred in the Soviet Union and Eastern Europe. They have made possible the ending of the post-war era of military-ideological confrontation, an ending marked by the signing of an agreement abolishing medium-range nuclear missiles in December 1987. These missiles are only a small part of the East–West armoury, it is true, but their political importance goes far beyond percentages, and not merely because the treaty eased the way for agreements on strategic weapons and conventional forces in Europe. Anyone who attended the ozone layer conference in London in March 1989 could see how the end of the cold war, and the easing of so many rivalries and points of conflict, was making itself felt in the tone and level of participation by all the parties, including those from the poor South. It is doubtful whether a conference of that kind would have been as successful if it had been held a few years earlier. Euphoria should be restrained, though. The old East–West interface may be fading from such occasions, but a new one, based on different orientations, can be observed between the industrialized North and the developing South.

Furthermore, the ending of the bipolar political world should not be assumed to mark the beginning of a new multipolar world, with Japan and Europe entering the ranks of the superpowers. Wealth will still mean influence, but military power in a world where poverty and not ideology marks the divide will be hard to wield effectively for national advantage on the world scale. In any case, climate change is profoundly non-ideological, democratic even. We are all, rich and poor, in the same boat.

The third reason for the pace of change is that the moral disquiet underlying so much ecological thinking has steadily become transformed into the everyday stuff of cabinet meetings and parliamentary debates. Ecology has been absorbed by politically charged environmentalism as surely as the thinking of Adam Smith and his followers became the intellectual underpinning of the modern market economy. We are all environmentalists now, we can declare, whether we are radical green or right-wing conservative. The widespread perception that climate change is real and will affect not only our lives but those of generations to come is responsible for the conversion from the notion that there is money in industrial muck to the belief that the muck might make the money meaningless. With that comment should go a caution that public concern about the dangers and imminence of climate change will fluctuate. Even with such a serious subject there can be a surfeit of dramatic warnings and eco-friendly items on the supermarket shelves. Repetition dulls. The volcano rumbles, but after a while people return to its foothills to till the fertile soil and soon they forget – at their peril – about the danger.

If there is one other premise on which this book is constructed, it is that we have a resilient tradition of international negotiation in which concern for people is uppermost. It has been with us since the St Petersburg Declaration on Exploding Bullets in 1868, and perhaps from even earlier: from the day in June 1859 when the Swiss evangelist Henri Dunant, founder of the Red Cross, walked over the battlefield at Solferino and saw the wounded not as the pitiable debris of war but as men who should be regarded as neutrals with a right to be helped and protected by rules agreed among the nations. Observance of the Hague Conventions' constraints on warfare may not have been much in evidence in two world wars, but the tradition of international negotiation has developed into a culture with its own institutions and fora. The League of Nations, founded after the First World War, became the much expanded and strengthened United Nations after the Second World War. Henri Dunant's Geneva became the seat of the superpowers' negotiations on nuclear weapons. And, with negotiations on nuclear weapons, the overlap into environmental agreements becomes apparent. Nuclear tests caused atmospheric pollution which spread far beyond the national boundaries of the testing countries. International as well as national concern in the United States led to a ban on atmospheric (as well as space and underwater) testing in 1963, one of several international military-environmental agreements ranging from the use of weather modification techniques to the stationing of weapons on the moon. It is not such a far cry from them to the intensive international effort which was put into completing the report of the Intergovernmental Panel on Climate Change in time for the Second World Climate Conference in October–November 1990. Next lies the convention on climate and its

attached protocols on the greenhouse gases; and then, the
· 1992 UN Conference on Environment and Development,
where we shall begin to see whether we can control and live
with our progeny, both industrial and human.

*

An introduction is a useful place for an author to outline his
approach and to explain what he means where he has
stretched the boundaries of categories. Diplomacy, in the
greenhouse era, extends well beyond foreign ministries.
Environment ministries have their own foreign affairs sections. Energy in particular becomes increasingly an international issue, its use and production key factors in our attempts
to balance economic growth and costly controls on greenhouse
and other emissions. The dilemma presented by nuclear
power and its attendant cost and dangers in the search for
'clean' energy remains unresolved at the beginning of the
1990s. All these are diplomatic issues in the wide sense
demanded in the search for environmental security. So are the
rainforests, but the most important environmental battles will
be fought against a background of smokestacks, nuclear
reactors and overcrowded highways. Diplomacy in such a field
is, of course, the usual patchwork of national interests stitched
together in the mutual interest of keeping out the chill, or, in
this case, the heat. Some see the national interest as served by
expanding their economies as fast as possible; others by
setting examples that will lead to international agreements on
the environment; and yet others take a middle road, reluctant
to take steps that will harm or hamper fragile economies until
they know beyond doubt what is causing climatic change.
Britain and the United States are among the last group.

The gurus of ecology may have signposted the now crowded
road, but it is the climatologists who lead us stumbling along
it. We all know enough to be alarmed and it provides the
context in which we interpret, rightly or wrongly, the visible
evidence of those old foes, drought and famine, and their
victims, withered trees and cadaverous bodies. Thanks to
television, they are part of our environmental scenery, a
disturbing horizon on our domestic screens which exists

beyond the green one we see from our windows. Pollution is no longer just pesticides which kill songbirds; it is rising sea levels with the potential to destroy countries.

The threat of a rapid global warming has changed the rules of the political game, and perhaps the name of the game as well. It has, as I have suggested, become environmentalism, with a wider and less refined set of rules than ecology, with its philosophical concern for resource conservation and holistic opposition to the physical sciences. When politicians have to worry about the effect on the climate of radiative fluxes from the sun they are apt to regard ecology's emphasis on biological science as a bit old-fashioned. The 'hard' scientists receive the sort of rapt attention from governments that is normally reserved for wartime. They have to work fast and think big – globally, in fact. No one can be a neutral and opt out of responsibility for global warming. It is not enough to make defensive agreements among allies. Everyone has to agree. Thus China's plans to burn more coal to produce more energy is not seen solely as welcome evidence of a great country moving into the industrial mainstream, but as a possibly dangerous addition to the tonnage of carbon dioxide received every year by the atmosphere.

And there one comes to that abrasive interface between North and South. The greenhouse gases in the atmosphere are overwhelmingly the by-product of the industrialized nations of the northern hemisphere's mid-latitudes – those latitudes where climate warming is expected to show some of its more extreme increases. It is not unrealistic to imagine arguments opening up on the lines that the big polluters should cut their emissions while the developing countries – notably China, India and Brazil – should be allowed to pollute more until a sort of balance of pollution is reached. And then, of course, there is the question of environmental aid to developing countries. 'Making the polluters pay' has applications which go beyond the cost of domestic electricity bills. Will the industrialized West pay the developing countries to be 'clean'?

*

In its formative years, ecology was the speciality of Germany,

Britain and the United States. The new environmentalism is widespread, but it is still essentially a Western phenomenon. Its manifestation in the form of green politics has been remarkable. In the European elections in June 1989 the British Greens took 15 per cent of the vote, more than the centrist Liberal Democrats and Social Democrats. But who are the Greens? They have no hard-and-fast political centre and, at times, seem more like an emotion in search of a solid platform than a real party. The British party suffers from a crippling shortage of funds because it is not concerned with the usual business of parties in a democracy: the protection of class interests or economic principles. The Greens represent fundamental concerns, even, if you like, instincts about the world and a betrayal of nature. Essentially, they are still lobbyists enjoying all the delicious freedom and righteousness that fringe opposition can bring; and the extremism, and even illogicality, of some of their factions can sometimes do their image a great deal of harm. It is best to see them as a broad movement which has brought together tendencies from right, left and centre. Green is already becoming as generic a political term as Red. The Greens may build parties which, given the changing conditions of the environment, will have real influence within the elected assemblies, or they may be absorbed by the general greenness which is colouring all political parties. I suspect it will be the latter, simply because government will increasingly become a balancing act between the interests of the environment and those of the economy, and balance of that sort is not yet perceived as a Green attribute.

*

Life in the West since the Second World War has been so good that it is hard to adjust to the view of the twenty-first century presented by the statisticians. If they are wrong about a doubling of the world's population, it will be because disaster has intervened on a scale more massive than either of the two world wars. If they are right, it is hard to see how African, Latin American and many Asian countries can avoid increasing poverty and political turbulence. How does a wretchedly poor country like Ethiopia deal with a tripling of its population during the next four decades? The fact that Western banks are being forced to write off their loans to Latin American countries is an indicator of how hard the task of development will be, even without the prospect of disruption caused by climate change. But gloom is a bad counsellor. The future of the world must have looked even bleaker in some respects on the eve of the Second World War. Since then the machinery of global security has become a great deal stronger and more sophisticated. We may not be able to avoid the disasters caused by a combination of global warming and population growth but we may be able to mitigate them. Or, to put things in a different way, the twenty-first century may not turn out to be the end of the world, but it could be a closer shave than many people would care to imagine.

CHAPTER ONE

Setting the Agenda

In March 1989 the world's first environmental summit met in The Hague. It was a curious, poorly conceived event and its significance lay in the fact that it happened, took place in a city synonymous with international arbitration and, through its shortcomings, illustrated some of the problems confronting the world as it begins a search for international agreements and mechanisms to control the emission of greenhouse gases. The invitations were restricted in number by the summit's sponsors – France, the Netherlands and Norway – and of the twenty-four countries that attended, not all sent their heads of government. The three greatest polluters – the United States, the Soviet Union and China – were not invited because it was felt that 'the great powers might adopt different positions on the subject at this stage'. The true reason why they were not invited was, said the cynics, that it was a French-inspired summit and the superpower leaders would have overshadowed President Mitterrand and his prime minister, Michel Rocard, both anxious to imprint on the minds of the French electorate, and no doubt the world at large, the impression that they were in the vanguard of green thinking. And in the event the United States and the Soviet Union were to demonstrate, only eight months later at a follow-up conference in Noordwijk (The Netherlands) that so far from taking different views, they took the same view, at least on carbon dioxide: neither was ready to agree on cuts in emissions. The real difference existed between them and their hosts.

In China's case there was perhaps better reason for exclusion. China is the 'wild card' in environmental diplomacy: archaic, disturbed, huge and self-engrossed. Its leaders are intent on a great leap forward in energy production, above all in coal output, as the cure for industrial backwardness. Between 1980 and 1986 China's carbon dioxide emissions rose faster than those of any other country, by 27 per cent, although per capita output was still small by comparison with the developed countries.[1] Its stated industrial plans include the manufacture of 50 million refrigerators using chlorofluorocarbon (CFC) refrigerants, which are both ozone-depleting and a greenhouse gas. Consequently, Beijing was regarded as unlikely to be receptive to proposals for enforceable international agreements restricting pollution. The suppression of the student movement for democracy in Tiananmen Square in May 1989 was to demonstrate how indifferent the Chinese leadership was, ultimately, to international opinion.

The leader of another Third World mega-state, President Sarney of Brazil, the destruction of whose rainforests causes more international outcry than any other environmental issue, sent a representative. A meeting which might produce proposals threatening Brazil's sovereignty over its resources could not be dignified by his presence, it was decided. And Britain's prime minister, Margaret Thatcher, who only a week before had held her own, successful, conference on protecting the ozone layer, declined her invitation on the grounds that she did not approve the summit's declared goal of creating a new UN organization to oversee a law of the atmosphere convention under the ultimate jurisdiction of the Hague International Court of Justice.

What emerged from the conference was milder than billed, like the warm winter and spring which had brought the Dutch capital's trees and crocuses out in early leaf and flower, a precocity that was generally attributed to climatic change. The summit declaration left open the question of whether the proposed atmospheric convention should be administered by a new UN organization or by a strengthened UN Environment Programme, a body created in 1972 and based in Nairobi. Earlier talk of sanctions against offenders was muffled by an agreement to promote 'appropriate measures' to encourage compliance, with adjudication in the hands of the International Court of Justice.

It was a disappointing conclusion in some ways for the host, the Dutch prime minister, Ruud Lubbers, for whom a major attraction of the summit had been the prospect of having the new environmental institution established in The Hague (a goal later abandoned). Within a month his was to become the first government to lose office on an environmental issue: his Liberal coalition partners rebelled against proposals to raise the tax on diesel fuel and end a tax concession to car commuters. It was a brief and well-rewarded martyrdom. A general election in September returned him to power with an increased number of seats for his Christian Democrats at the expense of the Liberals who had brought him down.

A culture of negotiation

The Hague summit and the Noordwijk conference may not have been resounding successes, but what better place than the Netherlands, a country ingeniously and defiantly created by pushing back the sea, to assemble the salient problems of the environmental crisis? The 1990s will show whether the globe can agree on action to contain the crisis, and the Hague's identification with civilized negotiation was one reason for choosing it as the summit's site.

A peace conference not only supposes common interests and values; it takes it for granted that there are potential and actual enemies liable to adopt cruder methods of settling disputes than sitting round a table. And if we look at the history of arms control and the attempts to civilize the battlefield, and relate its lessons to the environmental crisis, we have to work on the same premises. There are common interests, and there are indeed potential enemies – although governments do not care to say so: the minority of rich nations and the majority of poor ones. In between there are the semi-rich (or semi-poor) like the Soviet Union and the Eastern European countries, who would like to be richer and are not quite sure where they stand. There are fiercely defended rights, of course, like sovereignty, which is significant for developing countries whose rapidly expanding populations demand evidence that they are escaping poverty and are little concerned in the short term if that escape is achieved at a high cost to the environment. Rich countries too turn to sovereignty as a last resort when environmental pressures become intense. A climate convention backed by sanctions against violators would cause as many problems for someone of Mrs Thatcher's or President Bush's ideological disposition as it would for President Sarney's successor, Fernando Collor de Mello.

Environmental issues have been a part of political life in the West since the 1960s, but it was in 1988 and 1989 that they truly entered the political mainstream, deepening and broadening the channels pursued by practically every department of state. Awareness of the threat posed by atmospheric degradation gave green issues a new prominence in the minds of electorates. It was not, however, new for twentieth-century humanity to be faced with apocalyptic warnings. That had already happened in the late 1940s and early 1950s with the creation of nuclear armouries whose destructive force made it clear that a third world war would not leave much for the victors to celebrate. By the mid-1980s there were more than 50,000 warheads, a figure that gave credibility to those, like the US scientist Carl Sagan, who theorized that a war between the alliances of East and West would trigger a nuclear winter, ensuring that all who had not been killed by fire, blast or radiation would be starved or frozen to death. Two television plays on the catastrophic effect of a nuclear war raised the level of public concern to even greater heights, particularly in the United States.

Deterrence was based on MAD – mutually assured destruction. But that acceptance, and its black joke of an acronym, never touched the public imagination as vividly as nuclear winter, with its image of sunless wastelands under smoke-black clouds. It was more potent than any listing of megatonnages and the throw-weights of delivery systems, and it played its part in readying minds at every level and class for an eager acceptance of the negotiating breakthroughs that came in the second half of the 1980s. Re-reading Jonathan Schell's warning of impending doom, *The Fate of the Earth* (1982), one is struck not so much by his underestimate of humanity's instinct for survival and its ability to embrace concepts of security which go beyond military alliances, as by how little editing would be required to make it stand as a visionary account of the threat posed by climate change – an environmental MAD. The crisis caused by global warming has a more complex structure than nuclear confrontation, however, and in domestic-political terms it cannot be reduced to the level of 'whose finger on the button'. Everyone's finger is on the button; or, more precisely, on the air conditioning and central heating switches.

The difference between now and the last great environmental rumbling in the early 1970s is that today it is governments which are promoting awareness of a common danger, not private organizations like the Club of Rome, a leader in the field for a few years. Last time it was the reckless pursuit of economic growth and the depletion of resources by Western society which were seen as the danger; and the warning was promptly ridiculed as doomsday scaremongering. This time there is hard evidence – evidence that we can all observe – that something unwelcome is indeed taking place. Savage drought in North America in three years of the 1980s, gales and disruption of weather patterns in Western Europe, drought-induced famine in East Africa, continuous rain in normally dry parts of Australia have all helped convince us that this is not an alarm created by politically-minded ecologists or another example of the quasi-superstition which, in the 1950s, attributed disturbed weather to nuclear tests. The scientific evidence is to be found in the temperatures recorded by a multitude of weather stations scattered around the globe. There is a general acceptance by scientists that the 1980s was the hottest decade on record, that the volume of carbon dioxide in the atmosphere is steadily increasing, and that the ozone layer over the South Pole shrinks back like a membrane touched by acid during the Antarctic spring. Rising temperatures and sea levels fed by the soufflé effect of thermal expansion in the twenty-first century will threaten productive agricultural regions with ruin. Sea-level rises pose a threat too to most of the world's great cities, from New York to Tokyo, London to Buenos Aires, and Shanghai to Leningrad. It is a unique crisis and it calls for co-operation and agreement with an urgency and on a scale far beyond anything in human experience.

A marginal existence

'There will be no winners in this world of continuous change, only a globe full of losers,' warned Michael Oppenheimer, the Senior Scientist at the Environmental Defence Fund, in his evidence to the June 1988 US Senate hearings on climate change, an event which perhaps more than any other awoke the world to the gravity of the situation:

> Today's beneficiaries of change will be tomorrow's victims as any advantages of the new climate roll past them like a fast-moving wave. There will be a limited ability to adapt because our goals for adaptation will have to change continuously. The very concept of conservation on which environmentalism in this country was originally built does not exist in a world which may change so fast that ecosystems, which are slow to adjust, will wither and die.

This willingness on the part of scientists to pronounce on the potential statistical measurements of the next century is new. So is the sadness when we consider the combined effects of climate change and a projected increase in world population from five to ten billion. Will wars, famines and diseases lead to human tragedies in the twenty-first century on a scale unrivalled even in the twentieth? Climatology's subject matter is wayward in the extreme, but its practitioners can make broadly acceptable worst-case, best-case and middle-of-the-extremes case projections for climate change based on what we do or don't do to control pollution worldwide. The climatologists can look at the record of the past trapped in ice cores which show the fluctuations in temperature and atmospheric content during the past 12,000 years of our interglacial and compare it with present trends. And as interglacials are no more than interludes in the long march of the ice ages, we know that our civilization – some 4000 years old, if measured by written records – is the gift of the interglacial, just as life is the offspring of an atmosphere 99 per cent of which is contained in a layer no more than 20 miles deep.

In breathable living space we are not much better off than the fish. We lead the most marginal of existences, and awareness of our vulnerability increasingly shapes our thinking in a way that was once the unchallenged prerogative of religion. Civilization is not guaranteed continued progress, much less eternity. One can look at the archaeological evidence of societies afflicted by climatic change, in southern Greenland as well as on the fringes of the North African and Middle Eastern deserts, and ponder their suffering. And one can cast thoughts more widely to a universe increasingly illuminated by astronomers, physicists and mathematicians and take some comfort from its impartiality. It is neither benign, malign nor implacable, and the products of humanity, whether they be nuclear power or organic farms, are as much part of its natural processes as solar flares and the methane gas produced by termitaries. In an infinite universe nothing can be alien or imposed, even if they are not necessarily replicated endlessly. We are not doomed because we are defying the laws of nature; short of a natural cataclysm, we will be doomed only if we are reckless or ungovernable. If in response to what we perceive as our needs we have endangered our survival, then it is both possible and necessary to recognize the danger and accept the remedies which go with changed needs. Living on the earth is a balancing act: we will fall if we lose our nerve or are victims of suicidal impulses. We might perish anyway through a landslip or lack of concentration. But at least the post-war nuclear world has shown that our nerves are strong and our impulses controllable.

A political sea-change

When Mikhail Gorbachev and Ronald Reagan signed the Intermediate Nuclear Forces treaty it marked the end of an age in which the world had been dominated by rival political and economic systems and their nuclear weapons. There are still massive stockpiles of nuclear weapons, but they no longer dominate our thinking.

It was the collapse of the Soviet economy as much as recognition of the costly futility of nuclear competition with the United States that impelled Gorbachev to sign what was fundamentally a peace treaty. But there was more to it than that. Gorbachev himself noted that the ubiquity and all-pervasiveness of the modern information and communication network was a major change in itself. An authoritarian state could no longer regulate and censor information. In the Soviet Union and other Eastern bloc nations people knew that free market economies produced consumer goods, abundant food, houses and all the many things that were so absent from their system. Life was better on the other side of the wall (it was also less polluted, and that was a factor too): the standards of living and abundant products of the capitalist democracies set the levels of aspiration. Competition between East and West had been switched from armaments to consumer goods.

There is good news and possibly some bad news for the environment in this sea-change. The good news could be seen at the March 1989 and June 1990 ozone-layer conferences in London, the first attended by 120 countries from East, West and the Third World. They spoke a common language of concern and there was no publicly voiced dissent from the objectives of the conferences. The mood was positive. Only a few years earlier such near-unanimity would have been impossible. There would have been recriminations, jockeying for political advantage between East and West and a great deal of suspicion. The ending of the cold war had made success possible.

The possibly bad news is that satisfying long-repressed consumerist aspirations in the Soviet Union and its allies could generate even more pollution in countries where pollution is already deemed a serious threat to health. The citizens of

Eastern Europe have turned to democracy because they see its freedoms as productive of all the material things they have been denied. Balancing a fear of the political consequences of disappointment with the expensive task of clearing up the environment will be a formidable task.

In the second half of the 1980s, nearly 80 per cent of the world's wealth was in the hands of the 15 per cent of its population who live in the industrial market economies of the West and Japan.[2] And wealth and atmospheric pollution tend to go together. Nearly half the annual man-made carbon dioxide emissions come from the rich market economies. Governments are still feeling their way forward in the search for solutions, and attitudes tend to be cautious and based on 'Let's wait and see what the others do'. That, in turn, means waiting for the scientists, who themselves are often torn between the purist caution of their profession and a realistic assessment that decisions have to be made before all the evidence is in. 'We made a mistake with the ozone layer by being too careful,' a British scientist told the author. 'We don't want to repeat that mistake with the greenhouse effect.' On the other hand, there is a widespread feeling that James Hansen of the Goddard Institute for Space Studies jumped the gun when he told the June 1988 US Senate committee hearing that he was 'about 99 per cent confident' that the increase in global temperatures – and the devastating 1988 American drought – was due to the greenhouse effect.

Scientists now accept that there is a greenhouse effect, but they are still measuring it and they do not know what the regional consequences will be. Most agree, though, that it is better to be safe than sorry, even if being safe is expensive. If governments are to achieve the target set by the Toronto conference of a 20 per cent reduction in 1988 levels of carbon dioxide emissions by 2005 they need to move immediately. The 20 per cent cut is only a first step towards the 50 per cent reduction which the conference agreed would be needed to stabilize atmospheric concentrations of carbon dioxide. The Noordwijk meeting demonstrated how difficult getting firm international agreement will be. Britain, the United States, the Soviet Union, China and India (and, for different reasons, energy-efficient Japan, which is far ahead of other industrial nations) united against a draft agreement calling for the stabilization by the year 2000 of carbon dioxide emissions at 1988 levels, followed by a 20 per cent reduction over the following five years. There emerged the damp fudge which usually occurs when the warm winds of aspiration hit the cold front of cost and national interest. The meeting recognized the need to stabilize carbon dioxide emissions, probably by the year 2000; the feasibility of a 20 per cent reduction by 2005 was something that would be studied. Similarly, the July 1990 Houston summit of the seven leading industrial democracies avoided conflict over targets and limited itself to calling for a climate change convention by the end of 1992.

Pricing poverty

In the summer of 1989 an unpopular British Environment Secretary, Nicholas Ridley, was moved to make way for the Conservative government's most notable green, Chris Patten – a testimony in itself to the new power of green politics. One of Patten's first acts was to disturb the doldrums of August by promoting a neglected report by one of his mentors, Professor David Pearce, head of the London Environmental Economics Centre. His subject, dealt with in a 181-page report,[3] was sustainable development, defined by the UN World Commission on Environment and Development in *Our Common Future* (1987), as development that meets the needs of the present without compromising the ability of future generations to meet their own needs. It is the answer to the post-war ecologists who in the 1960s and early 1970s claimed that the industrial way of life was inherently unsustainable and growth had to be cut back. Pearce took *Our Common Future*'s definition a little further with this formulation:

> This generation should ensure that it passes on to the next generation a stock of assets no less than the stock it has inherited. We must learn to recognize that environmental capital is just as much capital as man-made

capital. Environmental capital includes not just the stock of oil and gas, coal and minerals. It also includes the ozone layer, the protective functions of forests and wetlands, the waste-assimilating functions of rivers and oceans, and the store of biological diversity.

Sustainable development means growth within the bounds set by the need to maintain 'critical' environmental capital. The environment cannot be left to market forces, which treats many environmental resources as if they have zero prices. To prevent these resources from being over-used and abused they must be priced.

Whether we can price the ozone layer or a balanced atmosphere seems doubtful, but we can cost environmental damage and make the polluter pay. Up to a point. It is relatively easy to impose extra taxes on 'dirty' fuels, 'dirty' smokestacks and cars whose fumes are above set levels of toxicity. Governments – and tax-payers – can pay incentives to encourage cleanliness. But it is not so easy to assess the global environmental damage caused by cutting down a rainforest and then impose payment of the 'cost' on the 'polluter'. How, short of sanctions, can the principle of making the polluter pay be imposed internationally? Only through agreements approved and implemented voluntarily by all parties, is the answer. And negotiating conditions of the polluter-pays sort would almost certainly complicate and lengthen the process of reaching agreement. One obvious alternative is to avoid giving aid to projects considered environmentally damaging and to direct it instead to projects which enhance environmental health. But that, of course, is an approach which contains the assumption that the industrialized democracies can reach agreement on standards, and adhere to them. The main problem is to be found in the developing countries, notably Brazil, India and China. Poverty, steadily increasing populations and environmental degradation are issues that intertwine dangerously with sovereign rights, including the rights of countries to set their own priorities. Buying out those rights where they conflict with the containment of atmospheric pollution will be expensive, possibly prohibitively so unless everyone agrees that the crisis is so universally threatening that self-interest dictates co-operation. The self-interest in doing so seems obvious, but self-interest is often a selfish variable.

These, of course, are the issues that will work their way through the Second World Climate Conference to the agenda of the 1992 UN Conference on Environment and Development. There have been proposals for a grand Law of the Environment, a framework convention on the atmosphere and a convention on climate change. They overlap and the ideas are sometimes blurred, as a meeting of international legal experts, convened in Ottawa in February 1989, discovered. By early 1990 attention focused mainly on a 'framework convention on climate'. The Ottawa lawyers thought that seven protocols should be attached to the framework, covering: carbon dioxide, methane, CFCs and halons (both ozone depleters), nitrous oxides (mainly fertilizers and fossil fuels), tropospheric ozone (smog, ground-level atmospheric pollution), deforestation and reforestation, and a world climate trust fund.

In the early and mid-1980s the developing countries found themselves paying more to service and repay their loans than they were receiving in new money. By the end of the 1980s it was estimated that the developing countries owed $1300 billion, a burden of such colossal proportions that many debtors were unable to do more than make part-payments to meet their obligations. Ivan L. Head, President of the International Development Research Centre, in Canada, wrote in the summer 1989 edition of *Foreign Affairs* that since 1970 debt-servicing payment had risen more than tenfold: 'What was once a condition of illiquidity in much of the South has now become a condition of insolvency.' The role of the World Bank had shifted from being a net provider to being a net taker. Some countries, though, claimed they had nothing left to take. Lloyds Bank estimated that of the twenty-nine countries which it considered to be in difficulty in the second half of 1989, more than half were no longer paying interest. In the same period, the four main British banks wrote off £2 billion of Third World debt and the United States came to the aid of its neighbour Mexico (burdened with a total external public and private debt of $100 billion), and several other notable debtors deemed politically important to the United States, with the so-called Brady plan.

The days of lavish lending to the Third World were in the 1970s. They were tailing off by the early 1980s. The World Commission on Environment and Development (the Brundtland Commission) estimated in *Our Common Future* that the projected increase in international capital flows to developing countries during the rest of the decade was only half what was needed to restore growth to a point where it would reduce poverty: 'the reduction of poverty itself is a precondition for environmentally sound development. And resource flows from rich to poor – flows improved both qualitatively and quantitatively – are a precondition for the eradication of poverty.'[4] If Western banks were having to write off billions in debts at the end of the 1980s, will rich countries be willing in the 1990s to provide the billions which will be needed to pay for preventing the destruction of rainforests and filtering the effluvia of Third World industrialization?

How much they will pay, and how willingly, and with what political and economic consequences to themselves, may turn out to be the most important questions that have to be answered in the last years of the century. If the advanced economies want global environmental agreements they will have to pay for them. Otherwise, the developing countries will give lip-service only to the principles and be unable to resist the pressure to provide energy and growth at any price for their swelling populations. Constraint may not be possible even then: or it could be an academic issue for the simple and appalling reason that economies and societies will collapse under the pressures. One only has to look at some of the national population projections to see that that is a strong possibility: Kenya's population is expected to almost quadruple between 1987 and 2025, when it will reach 83 million; in the same period, India's will grow from 798 to 1365 million and that of Bangladesh (a country about the size of England) from 106 to 217 million.[5] But let us assume that global agreements involving payments and technology transfers are made. Even before the tax-payers are finding the money to finance them, they will be paying higher electricity and water bills at home to finance stricter environmental standards.

'That will be the next challenge,' said the British Environment Secretary, Chris Patten. 'Will people be ready to translate their rhetoric and aspirations into cheques?'[6]

Notes

1. *Estimates of CO_2 Emissions from Fossil Fuel Burning and Cement Manufacturing*, May 1989. Carbon Dioxide Information Analysis Center, Environmental Sciences Division, Oak Ridge National Laboratory, Tenn., USA.
2. World Bank, *World Development Report* (Oxford University Press, 1989).
3. Published as *Blueprint for a Green Economy* (Earthscan Publications. London. 1989).
4. *Our Common Future:* Report of the World Commission on Environment and Development (Oxford University Press, 1987), p. 69.
5. *World Development Report*, op. cit.
6. Interview with author, 22 June 1989.

CHAPTER TWO

Ideals and realities

It is, of course, the politics of the environment that drives the diplomatic machinery; and the politics derives much of its power from the minutiae of grievance and concern. Like Edward Lorenz's much-quoted Amazonian butterfly which, by wafting its wings, sets off a chaotic escalation in climatic events that eventually produces a typhoon, the suburbanite who fights to save a nettlebed from the developer's bulldozer on the grounds that it is a sanctuary for tortoiseshells makes ripples that converge with others and become a profound swell of discontent. In much of the West people are torn between an idyllic rural ideal and a reality of crowded roads and unceasing construction which devours fields and woods and excretes hypermarkets, four-bedroomed executive homes and wider roads to take yet more vehicles. Despite the electronic revolution which brings so much of the world to the office or sitting room, modern society is constantly on the move. We demand more space to live in and move in, and the car is in many cases more essential to our sense of freedom than our right to vote, even though this mobility has qualitative limitations. City-dwellers who drive outbound on a Friday afternoon are likely to find themselves in a miles-long nose-to-tail crawl which probably produces more greenhouse gases and acid rain in the space of a few hours than many small states emit during a year. The greening of the media as well as their own senses will have made them aware that the world is imperilled. Species that have taken millions of years to evolve are becoming extinct at a rate which (according to some authorities, at any rate) has not been exceeded since the earth last ran into an asteroid belt. Yet we still want more cars, empty highways and detached houses set in acres of unblemished greenery. The green movement in its popular dimension is both enlightened and selfish, and often it is hard to disentangle the two.

The yearning for a better, greener world has been a commonplace of urban disillusionment for decades. Politically, this disenchantment has not produced any major changes in economic policy. No party has won a general election by refusing to build a six-lane motorway or by calling a halt to the expansion of cities and their suburbs. Fears of pollution have not stopped exploitation of the North Sea and Alaskan North Slope oilfields. Neither Chernobyl nor the ecological disaster caused by the oil spill in Valdez Bay (Alaska) in 1989 has led to any shifting of priorities away from satisfying industrial society's demand for energy. What they have done is strengthen the opposition in many countries to nuclear power and create a greater emphasis on standards and security, and that, of course, is one measure of the heightened awareness in most industrial communities of the dangers of reckless growth in a fragile world. Other measures indicate a shift in values, but usually contain their contradictory elements. For instance, a US survey showed that if forced to choose between the environment and the economy, 74 per cent would accept slower growth. But, while 70 per cent favoured stronger fuel efficiency standards, 66 per cent were opposed to a 50 cents-a-gallon tax on petrol, traditionally cheap in the United States.[1]

The environment and its protection now figure in all political platforms but disasters will still occur. They are part of our predicament. All industrial societies, whether capitalist or communist, live by the pulse of economic growth, and it is a dirty business. 'Sustainable' is the buzz-word in the West when attached to development, but it has created few images for PR practitioners beyond windmills for power generation and organic farming. 'Sustainable' ideas like fewer cars and more public transport, or more expensive electricity to promote efficiency, have not gripped imaginations yet.

Economic expectations remain as high as ever. Two related definitions of sustainable development were given in Chapter 1. Here is another, from Brian Mulroney, the Canadian prime minister, who formulated a painless version of the new credo in his address to the 1988 Toronto conference on climate change: 'We believe that there are no limits to economic growth, other than those inspired by our imagination, but we do recognize there are real limits to natural systems and resources.'

The previous notable surge of concern over the environment occurred in the early 1970s. It was deflated when governments and economists attacked the central premise that Western societies were greedily exploiting and squandering the globe's resources. The way in which the West overcame the 1974 oil shock by tapping hitherto unexploited oil and gas resources helped to discredit the environmental pessimists. Nevertheless, it remains true that oil reserves are being depleted at a rate which exceeds the discovery of new reserves. The prediction of the 1980 *Global 2000* report,[2] that oil production would peak in the 1990s and thereafter decline steadily for the next sixty or so years still holds up reasonably well.

Global 2000 was prepared for President Carter and promptly dumped by his successor, Ronald Reagan, who, like Margaret Thatcher, during most of the 1980s, tended to be dismissive of expressions of environmental concern, particularly when they contained an underlying note of foreboding. The study was intended as a sweeping reconnaissance of the future, the first and largest of its kind by a government. It covered environmental, resource and population stresses and came at the end of the period in which environmentalism was still largely a synonym for conservationism. Global warming was hardly mentioned. There were, said the report, 'unresolved problems' that made it difficult to say anything reliable about climate. This pre-greenhouse era dates from Rachel Carson's brilliant polemic against pesticides, *Silent Spring* (1962). With its image of a world in which a combination of greed and chemicals had destroyed the homely wildlife of the countryside, it brought awareness of environmental dangers to broad reaches of middle-class opinion which knew little of the philosophers of the environment such as Lewis Mumford and René Dubos.

The new ecological creed

Mumford, a New Yorker born at the end of the nineteenth century who died in January 1990, described himself as a 'specialist in utopias';[3] and his first book, published in 1922, was a study of ideal states, whose antithesis, Kakotopia, he was later to create to encapsulate the horrors of modern urbanization. The range of his thought went considerably beyond that of most modern Greens, but there are traces of his utopianism in the thinking of contemporary European Greens, with their dream of regional arcadias which renounce arms and supranationalism of the sort represented by Brussels, and live by bartering their produce:

> For its effective salvation mankind will need to undergo something like a spontaneous religious conversion: one that will replace the mechanical world picture with an organic world picture, and give to the human personality, as the highest known manifestations of life, the precedence it now gives to its machines and computers. This order of change is as hard for most people to conceive as was the change from the classic power complex of Imperial Rome to that of Christianity, or, later, from supernatural medieval Christianity to the machine-modelled ideology of the 17th century. But such changes have repeatedly occurred all through history; and under catastrophic pressure they may occur again.[4]

Men like Mumford were not prophets in the sense that they pronounced on the inevitability of doom. They were inspired publicists who preached conversion to the new ecological creed as the first step towards preventing catastrophe.

The post-World War II breed of eco-philosopher thus worked on well-prepared ground. They were helped, too, by the fact that the United States was by far the most advanced industrial society in the world. It had introduced the nuclear era; its industries and vehicles produced a greater volume and variety of pollutants and chemical poisons than anywhere else. The use of defoliants in Vietnam as well as Rachel Carson opened people's eyes to what chemicals were doing to nature. Saving trees had implications for the peace movement as well as for the opponents of acid rain, and very often they were the same people.

There is one other factor which is important: the United States is a litigious country and groups who feel their rights or their health have been impaired can combine to take class actions, a concept that remains alien to most other nations. It is not surprising, then, that the United States has often been ten years or more ahead of Europe in its application of environmental controls – on CFCs, vehicle emissions and chemical poisons. US public opinion has pushed harder and been more effective in exerting pressure on the political and legal machinery than in Europe until recently.

The teachings of the US ecologists have by no means been uniform. Dubos proclaimed that there was no 'natural' ecology. To understand what was happening it was essential to study humans, for they had changed everything. Professor Barry Commoner (who has never claimed the title ecologist and describes himself as a scientist) sees humanity in the grip of history, carried along in its march by a phalanx of technological, economic, social and political forces. The scientist can reveal the depth of the crisis, but only social action can resolve it. As a biologist, he believes that humanity has come to a turning-point in its habitation of the earth. Never before has the planet's shallow, life-supporting surface been subjected to such diverse, novel and potent agents. Scientists have a new duty in addition to their older responsibilities for scholarship and teaching; they must remove the restraints of secrecy on the environmental impact of new developments and products before they are introduced.[5]

Whereas Commoner sees the scientist as both creator and solver of the crisis, Paul Ehrlich, a biologist, attributed the crisis squarely to ordinary men and women, particularly those in the Third World. Ehrlich became increasingly concerned with the problems caused by population increases, turning the subject into a best-seller *The Population Bomb* (1968). Birth control programmes and the regulation of population were in those days a highly controversial subject. They were seen as essentially aimed at the Third World and ethnic minorities within Western nations, and therefore tinged with racism. That view has been tempered in recent years by the acknowledgement of the simple mathematics of population increases which outstrip growth in production and whose momentum will carry the surge forward into the second half of the next century.

What the American gurus had to say was on the whole fairly straightforward, but that could hardly be said of Teilhard de Chardin (1881–1955), a French Jesuit and palaeontologist, who coined the term noosphere (from the Greek *noos* for mind), to describe 'a particular biological entity such as has never before existed on earth – the growth, outside and above the biosphere, of an added planetary layer, an envelope of thinking substance.'[6] There are so many spheres of one sort and another in the realm of ecological, climatological and allied thinking that at times it has similarities to a medieval cosmologist's description of heaven, hell and the spiritual divisions of mankind, so perhaps a few definitions would be useful.

What we stand on is the lithosphere, the dead crust of the earth. Surrounding that is the biosphere, the terrestrial zone containing life. The concept of the biosphere comes from Lamarck (1744–1829), the French naturalist and evolutionist, its name was given it by Eduard Suess, an Austrian geologist and its modern gloss came from the Russian scientist Vladimir Venadsky, who published a monograph, *The Biosphere*, in 1929. Teilhard de Chardin gave the term his own definition of a 'vitalized substance enveloping the earth'. This biosphere was in turn surrounded by the noosphere, a sphere of consciousness in which a vast and ceaseless evolutionary process was taking place. Mankind, in his view, disposed of a 'vast reservoir of time' in which to achieve its evolution. Life had flourished quite paradoxically in improbable circumstances for 300 million years and no doubt would continue to do so: 'Does not this suggest that its advance may be sustained by *some sort of complicity on the part of the "blind" forces of the Universe* – that is to say, that it is inexorable?'[7]

Which brings us to the best-selling author and environmentalist, James Lovelock, a British scientist and academic, and founder of the eco-cult of Gaia, the living planet. Lovelock's hypothesis is that 'the biosphere is a self-regulating entity with the capacity to keep our planet healthy by controlling the chemical and physical environment'.[8] However, Gaia is not a synonym for the biosphere, Lovelock asserts in a companion work *The Ages of Gaia* (1988):

> Gaia . . . has continuity with the past back to the origins of life, and extends into the future as long as life persists. Gaia, as a total planetary being, has properties that are not necessarily discernible by just knowing individual species or populations of organisms living together.

Gaia is also a sort of earth goddess, interchangeable with the Virgin Mary: 'Gaia/Mary is of this Universe and, conceivably, a part of God. On Earth she is the source of life everlasting and is alive now.'[9] Gaia is central to the beliefs of the Deep Ecology movement whose slogan is 'Earth First' and whose followers (or some of them) believe that Mother Earth has no use for modern man.

The equilibrium state

It is, of course, the gurus, the television pundits, the journalists who set the terms, catch and focus a mood or an apprehension and bring it forward to the point where the politicians jump on board. Religion and pop-mysticism are part of the scene, but not an important part. In Britain and the United States, it was the scientists and the environmental pundits who provoked the debates that preceded the United Nations' Stockholm Conference on the Human Environment in 1972, a remarkable and seminal year.

The debate was opened in the United States by a controversial book, *The Limits to Growth*,[10] commissioned from the Massachusetts Institute of Technology by the Club of Rome and funded by the Volkswagen Foundation. A week before the publication of *Limits* in early March 1972 the Sunday edition of the *New York Times* gave its readers a chilling account of what to expect: 'A major computer study of world trends has concluded, as many have feared, that mankind probably faces an uncontrollable and disastrous collapse of its society within 100 years unless it moves speedily to establish a "global equilibrium" in which growth of population and industrial output are halted.' The task before humanity would require a 'Copernican revolution of the mind'.

The pessimism of the mood in some circles was typified in a statement by U Thant (UN Secretary-General 1962–71), which the editors used to preface *Limits*' introduction:

> I do not wish to seem overdramatic, but I can only conclude from the information that is available to me as Secretary-General that the members of the United Nations have perhaps ten years left in which to subordinate their ancient quarrels and launch a global partnership to curb the arms race, to improve the human environment, to defuse the population explosion, and to supply the required momentum to development efforts. If such a global partnership is not forged within the next decade, then I very much fear that the problems I have mentioned will have reached such staggering proportions that they will be beyond our capacity to control.

Limits' authors proposed the 'equilibrium state' as the answer to such foreboding. In such a state population growth and use of resources would be fixed. The birth-rate would equal the death-rate and capital investment would be the same as the depreciation rate. The desired levels would be in accordance with 'the values of society' and they would be subject to revision as the advance of technology offered new options. There were problems, the editors admitted, in making the transition from a growth society to an equilibrium state and the team's thoughts on the subject were not sufficiently developed to grasp all the implications.

Scepticism in the scientific community about the conclusions of the report and its use of figures for exponential growth (i.e. constant percentage growth) in its charts was immediate and blunt, ranging from 'utter nonsense' to 'simplistic'. Henry Wallich, a Yale economist and columnist for *Newsweek*, condemned *Limits* as a scare story: 'This is Malthus again,' he mocked, in a reference to the early nineteenth-century English economist who had forecast that population would outstrip resources. The idea of a no-growth world was 'a middle-class baby – they've got enough money and now they want a world for them to travel in and look at the poor.'

Limits' equivalent in Britain was *A Blueprint for Survival*,[11] written by Edward Goldsmith, editor of *The Ecologist*, and Robert Allen, then his deputy editor, and endorsed by thirty-three scientists, including the nearest Britain had to a scientist-guru, Sir Frank Fraser Darling.[12] Perhaps Goldsmith deserves guru status, too. He financed *The Ecologist*, which has been described as 'the intellectual core of the British ecological movement during the 60s and 70s'[13] and has campaigned hard politically and as a writer for environmental causes. The similarities of timing of publication and theme provide some justification for thinking that *A Blueprint for Survival* and *Limits* must have had a common source in the Club of Rome. Goldsmith denies this vigorously. 'It was convergent thinking,' he told the author in July 1989. 'I met Dennis Meadows [Director of the project at MIT] at a conference in Philadelphia in the autumn of '71 and he gave me a proof of *The Limits to Growth*, but we had already written *Blueprint for Survival* by then.'

Blueprint pulled no punches in its approach. It preached that indefinite growth could not be sustained by finite resources, and survival depended on halting growth and halving Britain's population:

> The principal defect of the industrial way of life with its ethos of expansion is that it is not sustainable. Its termination within the lifetime of someone born today is inevitable – unless it continues to be sustained for a while longer by an entrenched minority at the cost of imposing great suffering on the rest of mankind.

The British government was accused of concealing the realities of the situation and briefing scientists misleadingly. So serious was the environmental crisis, said *Blueprint*'s supporters, that the movement would, if necessary, form a political party and contest the next election. The prospect of being drawn into politics by a document in which some of the analysis was regarded as flawed and the statistics suspect was enough to frighten a number of the scientists who had endorsed *Blueprint*, but the document did sow the seeds of the present Green Party in Britain. (It began in 1972 as People, became the Ecology Party a few years later, and changed to its present name in 1985.)

The man who marched at the head of the counter-attack was not, as might have been expected, a politician or the chairman of the Confederation of British Industries, but another editor, John Maddox, of the internationally respected magazine *Nature*. He denounced *Blueprint* in editorials and in the national press for fanning public anxiety about problems that were either exaggerated or non-existent. Rachel Carson and Paul Ehrlich were also pummelled in the course of a vigorous polemic against what until then had been 'the American phenomenon of prophecies of calamity'. Before the year was out the hyper-energetic Maddox had produced *The Doomsday Syndrome*,[14] which he insisted was not a scholarly work but a complaint. Sandwich-board men proclaiming the end of the world was nigh had been replaced by a throng of scientists, philosophers and politicians. Famine was an 'unreal scarecrow' and much the same could be said of warnings about vanishing mineral resource. The test was not to keep science and technology at bay but to control them and ensure that they did not become dehumanizing or even misleading influences.

Stockholm: 'a turning point in history'

Was there or was there not an impending catastrophe? If there was, would it occur in the next century or would the world escape it thanks to proven scientific and technological ingenuity? However the dangers were measured, it was clear that there were serious problems ahead and the time had come to set up an international framework which could look at the evidence and co-ordinate action. The 1972 Stockholm Conference was, from any perspective, an event of great significance, and Kurt Waldheim, the UN Secretary-General, did not exaggerate when he described it, in a speech at Fordham University, New York, in June that year, as 'a turning point in history, when a major correction was introduced in the process of the industrial revolution'.

Among the 112 countries whose representatives gathered in the Folkets Hus, the Swedish trade union centre, poor countries outnumbered those which the industrial revolution had made rich by two to one. The conference logo (adopted by its offspring, the UN Environmental Programme) was a figure with arms outstretched to span the globe on which it was superimposed, and underneath was the slogan *Only One Earth*.

Despite the official emphasis on harmony, the conference carried the ideological burden of the time in the shape of a row over East German participation. A successful Western objection to East Germany's presence led to the Soviet Union boycotting the conference in support of its ally. China, the self-appointed champion of the Third World, denounced the 'imperialistic superpowers' for being the chief sources of global pollution and demanded compensation for the poor nations. And the Indian Prime Minister, Indira Gandhi, condemned the advanced nations for having secured their wealth through 'sheer ruthlessness, undisturbed by feelings of compassion or by abstract theories of freedom, equality or justice'. Predictably, the conference's secretary-general, Maurice Strong, a Canadian, who had been given the UN's new environmental portfolio, took the middle road. He opened the conference with a speech which called for economic activity to take place in 'dynamic harmony with the natural order'. He did not believe that societies could cease to grow; no growth was not a viable policy. At the same time, the situation demanded new concepts of sovereignty, new codes of international law and new international means of managing the oceans. A new liberation movement had been launched to free humanity from thraldom to environmental perils of its own making.

For those not admitted to the conference there was the Environment Forum and one or two smaller assemblies where more radical views could be expressed. This large and voluble fringe was housed in a tent city on a disused airfield south of Stockholm. A 'Celebration of the Whale' was held and the celebrants marched into Stockholm to demand a ban on whaling. The Olympic Games of Pollution made mock awards to firms with 'outstanding accomplishments in pollution' and a Swedish youth group, Pow Wow, organized tours to show delegates the seamier side of super-clean Stockholm.

Stockholm had been in preparation for two years, and like so many conferences, the event was largely publicity for what had already been decided in principle. In retrospect, wrote one observer reflecting on it a decade later,

> the primary accomplishment of the Stockholm Conference was the identification and legitimization of the biosphere as an object of national and international policy. Its resolutions provided standards for environment-related acts of government, which even regimes indifferent to environmental values felt obliged to acknowledge . . .[15]

The conference agreed to the creation of a new institution, the UN Environment Programme (UNEP), which on Third World insistence was based in Nairobi; the setting-up of the extensive Earthwatch network to monitor the quality of the environment; and 109 Recommendations for Action and the Declaration on the Human Environment. As a US State Department paper[15] prepared in advance of the conference noted, the environment was a 'new dimension of activity' for the UN, which had been constructed after the war on sectoral lines similar to those of most governments – health, transport, agriculture, money and so on. The environment was different because it cut across so many sectors. The dawning of international awareness that pollution was a global problem was, however, attributable to the discovery that fall-out from nuclear tests had led to significantly increased levels of strontium-90 in 'such homely essentials as milk and vegetables in areas of the world remote from the scene of any such tests'.

The body that steered international awareness towards the 1963 Partial Test Ban Treaty was the UN Scientific Committee on the Effect of Radiation (UNSCEAR), established in 1955. The treaty halted atmospheric testing by the United States, Soviet Union and Britain, but France and China, which did not sign, continued atmospheric testing until 1974 and 1980, respectively. That quarter-century period exemplifies how slowly the imperatives of defence were overcome by those of health. There may be a warning here of the difficulties the world will face in accepting agreements in which the imperatives of economic growth are made subordinate to the needs of a vulnerable environment.

Notes

1. Survey published by Cambridge Energy Research Associates reported in *Financial Times*, 8 February 1990.
2. *The Global 2000 Report to the President: Entering the 21st Century.* Council on Environmental Quality and Department of State, 1980 (published in UK by Penguin Books, 1982).
3. Anne Chisholm, *Philosophers of the Earth: Conversations with Ecologists* (Sidgwick & Jackson, London, 1972), p. 4.
4. Lewis Mumford, *The Myth of the Machine*, vol. 2, *The Pentagon of Power* (Secker & Warburg, London, 1971), p. 413.
5. See Commoner's *Science and Survival* (Gollancz, London, 1966), especially chapter VII, for the development of these arguments.
6. Pierre Teilhard de Chardin, *The Future of Man* (Collins, London, 1964), p. 157.
7. Ibid., p. 71.
8. J.E. Lovelock, *Gaia: A New Look at Life on Earth* (Oxford University Press, London, 1987 edn), p. xii.
9. J.E. Lovelock, *The Ages of Gaia: A Biography of Our Living Earth* (Oxford University Press, London, 1988), p. 206.
10. Donella H. Meadows, Dennis Meadows, Jörgen Randers and William W. Behrens III, *The Limits to Growth: Report for the Club of Rome's Project on the Predicament of Mankind* (Universe Books, New York, 1972).

11. *A Blueprint for Survival* (Penguin Books, London, 1972). There is an echo of Goldsmith and Allen's title in David Pearce's 1989 report on sustainable development, *Blueprint for a Green Economy*.
12. Sir Frank Fraser Darling. Vice-President of the Conservation Foundation in Washington, DC, and deliverer of the BBCs 1969 Reith lectures, which were entitled *Wilderness and Plenty* and called on man to change his attitude to nature.
13. Anna Bramwell, *Ecology in the 20th Century: A History* (Yale University Press, New Haven and London, 1989).
14. John Maddox, *The Doomsday Syndrome* (Macmillan, London, 1972).
15. Lynton K. Caldwell, *International Environmental Policy: Emergence and Dimensions* (Duke University Press, Durham, N. Carolina, 1984), p. 53.
16. *Documents for the U.N. Conference on the Human Environment, 1972* (Department of State, Washington, DC).

CHAPTER THREE

Climatology, survival and politics

The possibility that global warming caused by increasing amounts of carbon dioxide in the atmosphere would melt the Antarctic ice-caps had been given a serious airing, in 1965, by the US President's Science Advisory Committee. Seven years later, in May 1972, shortly before the Stockholm conference, Lester Machta, of the National Oceanographic and Atmospheric Agency, warned his audience at a Long Island symposium on 'Carbon and Atmosphere' not to be misled by the cooling observed over the past twenty-five years. It might have been the result of increased cloud-cover caused by volcanic activity. The use of fossil fuels, he warned, could mean an increase of carbon dioxide in the atmosphere sufficient to raise the world temperature by 1°F by the year 2000.

Speculation on those lines sounded new, but in fact greenhouse theory dates from the nineteenth century. John Tyndall, the Irish physicist, noted in 1872 (in *Contributions to Molecular Physics in the Domain of Radiant Heat*) that a comparatively slight change in the 'variable constituents' of the atmosphere 'would produce changes of climate as great as those which the discoveries of geology revealed'. Svante Arrhenius, the Swedish chemist, linked the possibility of global warming to increases of carbon dioxide caused by pollution; but it was not until 1938 that the British air pollution expert, G.S. Callendar, produced what is recognized as the first firm warning that burning fossil fuels would lead to world-wide climate change: 'I mark Callendar's 1938 paper as the point at which climate began to move centre-stage in world affairs,' said Kenneth Hare,[1] Chairman of the Canadian Climate Programme Board, at the 1988 Toronto conference on climate change. It was a slow movement despite the long history of greenhouse theory. The idea that industrial and scientific achievements had set in motion an insidious process which would put man on the defensive against a coalition of nature and his own works did not become hard political currency until the second half of the 1980s.

Even now, when there is firm evidence of global warming, there is a question-mark over the responsibility carried by man-made carbon dioxide. Could there be other natural causes, such as changes in solar radiation? Climatology is an inexact science conducted by a surprisingly small number of people – some 300 – regarded as experts in the field. Understandably, they allow wide margins for error in their predictions and avoid providing guarantees. The high degree of uncertainty is reflected in the reluctance of many governments – Britain, the United States and the Soviet Union among them – to adopt expensive counter-measures. And that reluctance in turn slows down the pace of environmental diplomacy. Economic self-interest is a factor, of course, but if one wants to understand why progress is so often sluggish, it is necessary first to take a look at climatology and the slow formation of ideas about climate and its impact on human society.

Perhaps what we have found difficult to accept is not so much the idea that the earth's climate is subject to violent change but that it could happen in the space of a lifetime. After all, we have known that the world was once very much colder since 1837, when Louis Agassiz, the renowned Swiss glaciologist, enunciated the theory of the Ice Age. In its day, it had almost as radical an impact on established thinking as Darwin's theory of evolution and you need look no further for the beginning of an intellectual acclimatization to the fact that the world's climatological calendar was long and man's appearance in it brief. No longer was it the Noachian flood that had covered the earth, it was ice. But however you regarded the ice ages, it was not regarded as even remotely likely that the next one might be just round the corner. A few thousand years away, perhaps, but not something to worry about with visions of the glacial front crunching its way through the artefacts of civilization, to regain its old frontiers across Europe and North America.

There is a modern theory of mass (but not total) extinctions every 26 million years, the last 11 million years ago, but these have been speculatively linked to catastrophes caused when the earth periodically passes through a belt of asteroids. An ice age, by contrast, would be a return to the climatic regime which has prevailed for 90 per cent of the past one million years, and in that sense only incidentally a disaster for humanity, which would survive in reduced circumstances. Parts of the world that have been covered by the seas since the start of the interglacial would once again become dry land as the ice-caps drew more and more water into their expansion. Atlantis might re-emerge in the Atlantic to confound those who claim it is Santorini in the Cyclades.

There were climatologists who theorized during the 1960s and 1970s that the cooling observed then was the next ice age breathing frostily down our necks. The periodic changes that produced ice ages were explained in the 1920s by a Yugoslav mathematician, Milutin Milankovitch, as the result of cyclical tilts in the earth's axis relative to the sun within a frame of 100,000 years. The northern and southern winters grow warmer and the summers cooler and the net effect is that more snow falls and more ice forms and stays. Ice increases the albedo (or reflectivity of the earth), so solar heat which would otherwise be absorbed is reflected back into the atmosphere. The last ice age was only 4°C colder than it is today and Soviet scientists have argued that the greenhouse effect might be beneficial as a counteraction against a new ice age. A Belgian and an American scientist, who produced studies based on Milankovitch's theories, were thinking along similar lines in their prediction that the fall in temperature towards a new ice age was likely to steepen in the next millennium. In the natural order of things, a modest 'glacial climax' might be 'only' 3000–7000 years away and a full one 60,000 years hence.[2] The probability of the onset of an ice age within the next century is rated at only 1 or 2 per cent, but if it did arrive (and was not offset by the greenhouse effect), research on the pollen record preserved in peat bogs suggests that its stages could be relatively abrupt and cause some sharp changes in vegetation within the space of as little as twenty years.

Hot and cold in the interglacial

Climatologists frequently point out that the popular idea that there is an underlying constancy in the climate is mistaken. There have been two pronounced swings away from the interglacial norm in the past 1500 years (and others before that). There are no definite beginnings and endings to these deviations, and the dates given by climatologists vary considerably. The 'Medieval Warm Epoch' appears, despite its name, to have begun as early as AD 600, reached a peak some 300 years later, and then declined erratically until petering out in 1400. The shift in the average climate seems small: 0.7–1°C warmer than the twentieth-century average in England and 1–1.4°C in Central Europe.[3] The 'Little Ice Age' which followed lasted from *c.* 1400 to 1800 and is commemorated in the British imagination by scenes of ox-roasting on the frozen Thames during the coldest snaps. Within that period there were considerable fluctuations, but the mean temperature in the late seventeenth century was about 0.9°C lower than the mid-twentieth-century average in England, with the difference widening to 1.5°C in the exceptionally harsh years 1690–9.[4]

The key question is: what effect did they have on the course of civilization's development? The answer has to be very little indeed, although regional impacts were severe. If one goes back to the sub-Boreal climatic period (3500–800 BC) there is a case for attributing the growth of the Indus Valley civilization to reliable summer monsoon rains, and its decline to their failure. Nearer home, the rise and fall in the levels of settlements on Dartmoor has been interpreted as an indicator of changes in climate and temperature.[5] Between 4000 and

2000 BC, the moor was cultivated up to 455 metres. A deterioration in the climate up to 500 BC led to the abandonment of land above 300 metres. With the return of warmer weather between 800 and 1000 AD cultivation reached to nearly 400 metres, then fell back to 300 metres during the Little Ice Age. The same climatic swings may have had a stimulating effect on the neolithic cultures whose stone temples and celestial observatories in Northern Europe testify to a growing awareness of the changeable and often hostile world in which they lived. In more recent times, the collapse of the Norse settlements in Greenland is usually explained by the Little Ice Age. In a marginal region even a small change in temperature can have a savage effect, as it did in Iceland, where ice cut sea communications, glaciers overwhelmed settlements and life was so harsh that the evacuation of the entire populaton was seriously considered in 1784.

The Norse settlements established in southern Greenland in the tenth century may have vanished by 1500, but the Inuits who were driven *into* the same region by a deteriorating climate seem to have survived. Their culture had been shaped by the rigours of the climate, and was consequently primitive. The Norsemen, on the other hand, had brought their culture with them from slightly more benign circumstances. As with the polar expeditions of the first half or the nineteenth century, there was little inclination to learn from the locals. Woollen clothes were modelled on European styles and their boats remained the long, clinker-built kind that had brought them to Greenland. Much effort was devoted to religion. Church bells and stained glass were imported and a cathedral built. One writer has suggested that the Norse settlements perished as a result of a 'fatal inflexibility'. A society which believes that 'lighting more candles to St Nicholas will have as much (or more) impact on the spring seal hunt as more and better boats is a society in serious trouble.'[6]

If there is a lesson to be learnt from Greenland it is that a culture which in one set of circumstances mobilizes and directs efforts can in another become an encumbering harness which drags society down. The Norsemen of Greenland did not imitate the ways of the despised *Skraelings* and become indistinguishable from them; instead they perished. Societies find it hard to change gear, lower expectations and accept different ways and lower standards of living.

A science searching for rules

There are no isolated Greenlands in the modern world. It has few terrestrial secrets and no sanctuaries from its major developments. Satellites monitor crops, deserts, weather, wars, urbanization and the cryosphere of ice and snow which includes the polar caps and the glaciers. Instruments soon detect an earthquake, or a nuclear test, or a disaster in a nuclear power station. With climate, which embraces so many factors, including the greenhouse effect, much can be measured, but no one is quite sure how the sum is made or, indeed, what it will eventually total. Climatology is a science which is still groping for rules. Until those are known with some confidence and incorporated in mathematical models, predicting the course of change will remain, to an uncomfortable extent, a matter for intelligent guesswork:

> From the study of both direct and indirect (or proxy) data of the past we know that many regions of the Earth have witnessed a long and complex series of varied climates. How this evidence fits together remains an unsolved puzzle, but like all other phenomena in nature, the climate and its changes are presumably governed by physical laws; the discernment of the laws is the goal of modern climate research. That this research is not yet complete is shown by the fact that there is at the present time no unifying general theory of climate, and it is therefore not surprising that there is great uncertainty over the prediction of climate change.[7]

Because the ocean changed much more slowly than the atmosphere, and the ice and land masses more slowly still, it appeared that the climate was never able to reach equilibrium. The changes were so gradual as to be almost imperceptible, but over hundreds of thousands of years they could produce climates as different as the present summer and winter. The 'spectrum of climate change' was made even broader by the influence of factors outside the atmosphere and oceans, such as changes in the sun's radiation and in the distribution of the earth's oceans and continents (which tend to move about) over geological times. In short, the same author was asserting, it was going to be very hard to find consistent patterns and rules in past behaviour that could be related to the new factor of industrially-induced climate change.

The lithosphere, or land area, plays the least active part in the puzzle; the atmosphere is the most important part; and the hydrosphere (the oceans and seas) which covers two-thirds of the planet's surface comes next (some say it is equal with the lithosphere), most of the solar radiation reaching the earth falls on it and is absorbed by its surface down to a depth of a few metres. The hydrosphere is largely unexplored and the part it plays in the overall climatic scheme is the subject of an ever-increasing research effort. The cryosphere fluctuates in size considerably according to the seasons. The Antarctic ice-pack, for instance, extends to nearly 60°S in winter, an expansion which, if applied to the northern hemisphere, would bring the ice-pack down to the Shetland Isles. Lithosphere, atmosphere, hydrosphere and cryosphere all interact with one another. And so, of course, does the fifth party: the biosphere of people, other animals and vegetation.

Perhaps it simplifies matters to make a distinction between grand climate – the huge scenario of climatic events stretching over hundreds of millions of years – and the climatic system which we know, the one that governs our interglacial and, at the parochial level, our weather. The mechanism of this system which circulates the atmosphere and within which the highs and lows of weather, grow, drift and disappear, is powered by the earth's surface heating and cooling. The sun delivers nearly two-and-a-half times as much radiant energy to the equatorial regions as to the poles. From the high pressure zones of the tropics to the polar lows there is a pressure gradient which brings air sliding polewards and drifting (to the right in the northern hemisphere) at the same time in obedience to the tug of the spinning earth. There are predictable ocean currents and winds which the navigators of sailing ships rely on when plotting their courses. The westerlies in the northern hemisphere blow in a north-east direction from the sub-tropical highs lying along the 30° latitude. Between them and the equator the trade winds blow into the doldrums in the low formed by the thermal equator, the line of maximum heat, which meanders north and south of the geographical equator. At the extremes of the globe are the great circumpolar vortices, the key elements in the system.

Grand climate may so far be fairly inscrutable, but the climate system is humble enough to bear a resemblance to the law. There is an overall design; one can understand the principles and logic that govern the mechanism. But within its rules nothing is fixed but is subject to radical change as the forces that shape and change the rules shift.

In climate the first 'principle' is the sun's radiation. Its heat is increasing, but so slowly it makes no difference to us and appears constant. Yet there are fluctuations and there is strong evidence from seventeenth-century records of a correlation between an absence of sunspots and the coldest periods of the Little Ice Age. Moreover, sunspots – regions of relatively low temperature accompanied by magnetic fields which sometimes throw off violent solar flares – are subject to cycles. And if cyclical changes in solar radiation can affect the weather/climate, could there be cyclical changes in cosmic radiation from the galaxy which affect the solar system? Is there such a thing as 'cosmos-climate' to be considered? No one has any answers on that, but the weight of scientific opinion has shifted towards acceptance that variations in the sun's activity affect temperature, weather and perhaps climate.[8]

It has been recognized for some time that there are eleven-year cycles of sunspot activity, but what has emerged recently is that there is a dominant 200-year cycle, and several other cycles as well. At the end of the 1980s solar activity was at a peak and scientists waited to see whether there could be a repetition of the pattern of 200 years ago, when the sunspots and warmth at the end of the eighteenth century were followed by very cold weather at the beginning of the nineteenth. If there were a repetition, would it offset the greenhouse effect?

The engine that drives the earth's atmospheric circulation

may be of simple design, but the circumstances that govern it are exceptionally complex. The cyclical tilts in the earth's axis relative to the sun appear to trigger ice ages; and the sunspot cycles presumably affect the highs and lows of temperatures in the ice ages as well as in the interglacials. Volcanic eruptions appear to come in cycles too, and may be subject to the tidal influence of sun and moon, and dust from them can cloud the atmosphere and cause cooling.

About 30 per cent of the sun's radiation falling on the outer atmosphere reaches the earth's surface directly through clear skies, and about 15 per cent is diffused through the clouds. The rest is absorbed or reflected by clouds, scattered and returned to space or lost to the ozone layer. That, at least, is the rule of thumb in the textbooks on meterology. The Intergovernmental Panel on Climate Change (IPCC) has experienced difficulties with clouds and their effects on temperatures and precipitation. To make matters more complicated, some clouds are more reflective than others. Tom Wigley, head of the University of East Anglia's Climatic Research Unit, theorized in *Nature*[9] that the polluted clouds which carry acid rain might actually be beneficial in lowering the rising temperatures caused by the greenhouse effect. The amount of sulphur dioxide (the main ingredient of acid rain) pumped into the atmosphere by power stations and other sources has increased sixfold in the twentieth century and 90 per cent is produced in the northern hemisphere, which, possibly because of this form of pollution, has not warmed as much as the southern hemisphere. The sulphate particles act as nuclei on which rain drops grow. More clouds mean that more solar radiation is reflected back into space, and Wigley speculated that acid rain clouds might have cut the potential greenhouse warming during the twentieth century by almost half. If Wigley is right, should the efforts to scrub power station emissions clean of sulphur dioxide be stopped despite the effect acid rain has on forests, lakes and human health? It would be ironic, wrote Wigley, if a successful clean-up accelerated the warming caused by the greenhouse effect.

At ground and ocean level the questions to be answered by the climatologists multiply. If large sections of the cryosphere of snow and ice began to melt and break up into the sea, there would be an effect on ocean temperatures and, since the reflectivity of the cryosphere would be reduced, more solar radiation would be absorbed by the earth, so increasing temperatures. The oceans are the biggest imponderable at the moment. Water is a poor conductor of heat and the slow mixing of water at various depths of the oceans through the currents and upwellings and the conflict of waters of different salinity means that it can take forty years for a heat exchange from the atmosphere to work though into ocean temperatures.

The Intergovernmental Panel

The difficulties of putting together a comprehensive picture of climate change which could be presented to politicians as a basis for action soon became apparent to the scientists and experts of the IPCC, established in November 1988 and given the Second World Climate Conference in autumn 1990, in Geneva, as the deadline for their report. The panel has a joint secretariat set up between the World Meteorological Organization and the UN Environment Programme. On the suggestion of Mostafa Tolba, the Executive Director of UNEP, three working groups were created: group one, headed by John Houghton, of the British Meteorological Office, deals with the scientific assessment; group two, headed by Professor Izrael, of the Soviet Union, covers the socioeconomic impacts of climate change; and group three, led by a US State Department official, Fred Bernthal, has to present a strategy for handling the crisis.

The IPCC can draw on the output of five computer climate models, four in the United States and one in the United Kingdom's Meteorological Office. In late 1989 the models were in agreement on the overall picture. Where they were in disagreement was on the regional impacts. A diagnostic unit was set up to see if it could sort out the reasons for the discrepancies and produce a unified picture. The part played by the oceans was one problem; another was the uncertainty about cloud cover. Would there be more or less clouds? Where and to what extent would they shield the biosphere from the sun's heat, and where and to what extent would they retain it? 'You need a fundamental breakthrough in cloud physics to solve the cloud problem,' said the panel's Secretary, 'Sam' Sundararaman, of the WMO, in the late summer of 1989.[10] Generally, though, it seems as if clouds have a net cooling effect. Solutions to the sort of problems facing the panel are not reached overnight, and so well before its report was published in the late summer of 1990 it had become accepted that it would not break new scientific ground, but merely establish what was known or had found a respectable consensus and put it in a frame of possible consequences and reactions.

'My feeling is that uncertainty will always remain,' Syukuru Manabe, of the US National Oceanic and Atmospheric Administration, was reported as saying in late 1989. 'We have to make decisions based on uncertain information. I don't think we have any other choice.' Two other US authorities on climate, Richard Lindz and Jerome Nahias, both members of the National Academy of Science, went further. They wrote to President Bush in late September 1989 to tell him that the predictions of climate warming were so inaccurate and uncertain as to be useless to policy-makers.[11] The George C. Marshall Institute, Washington, added to the doubts with a report which suggested that there had been a closer correlation in the past century between temperature changes and solar activity than with man-made greenhouse gases. The report produced in June 1990 by the IPCC's scientific working group did little to diminish the uncertainty. It found that while human activities were increasing the atmospheric concentrations of greenhouse gases, the increase in the mean global temperature over the past century was within the range of what might be expected from natural climate variability. The 'unequivocal detection' of the culprit was still a decade or more away. The report nevertheless spurred the hitherto fence-sitting British into pledging in May 1990 that Britain's carbon dioxide emissions would be stabilized at current levels by 2005. Attempts by the British and West German to persuade the United States to agree on a target for cuts at the July 1990 Houston summit of the Group of Seven failed, however, and there was no joint commitment. Equally unsuccessful was an attempt by the Europeans to persuade the

Bush administration to change its mind and support the idea of a World Bank global environment fund to help developing countries.

Climatology has become a key science, but its findings take time to work through. The scientists warn that even when they have managed to trim away some of the qualifications, a new generation of supercomputers may reveal more complex relationships that will take more time to work out. The many uncertainties associated with the magnitude, timing and effect of climate change made it difficult to know when to start doing something, the Environmental Protection Agency in Washington admitted in its February 1989 report to Congress on policies for stabilizing the climate. It might be prudent to delay action until there was more certainty about what was happening. The potential benefits of delay, however, had to be balanced against the potential increased risks.

But supposing the politicians decide on boldness and agree there should be immediate action, how quickly will it be possible to cut greenhouse gas emissions? There are promising technologies that could lower emissions, but they need more research and development to become 'economically competitive', according to EPA. In any case, nothing will happen quickly. It could take 20–50 years to introduce changes that would make a substantial difference to emissions. Science may be slow, but industry, it seems, can be even slower, particularly if its innovations have to be competitive.

Like military defence, environmental security and its related studies, industries and bureaucracies have become big business. No one knows how many billions of dollars are being spent worldwide on climate studies and monitoring, just as no one knows how many conferences on the subject of climate change, climatology and the environment are being held. The World Meteorological Organization declined even attempting to draw up a calendar. Governments as well as institutions are involved these days, since it is politically expedient to show a caring attitude towards the environment. So many conferences are a measurement of interest, but they can also be a form of procrastination, a constant re-treading of ideas and facts while international action is delayed in the interests of an elusive unanimity.

Notes

1. Dr F. Kenneth Hare, *The Global Greenhouse Effect*, Conference on the Changing Atmosphere: Implications for Global Security, Toronto, June 1988.
2. See H.H. Lamb, *Climate, History and the Modern World* (Methuen, London, 1982), p. 349, for a review of papers on Milankovitch's theory.
3. Ibid., p. 170.
4. Ibid., p. 201.
5. See Guy Beresford in Catherine Delano Smith and Martin, *Consequences of Climatic Change* (Dept of Geography, University of Nottingham, 1981).
6. For the fate of the Greenland settlements see Thomas H. McGovern, T.M.L. Wigley, M.J. Ingram and G. Farmer (eds), *Climate and History: Studies in past climates and their impact on man* (Cambridge University Press, Cambridge, 1981).
7. M.E. Schlesinger (ed.), *Physically-based Modelling and Simulation of Climate and Climatic Change*, Part I. Nato Advanced Science Institute Series (Kluwer Academic Publishers, Dordrecht, 1988), introduction.
8. *New Scientist*, 25 March 1989, and Report of UN conference on *Human Ecology and Climatology*, Leningrad, 1986. Sunspot activity can be monitored by examining tree rings from that time. The rings narrow during years when harsh weather restricts growth. After a few years an increase in carbon-14 atoms in the rings indicates that sunspot activity was at a minimum during the cold period. Carbon-14 is caused by the bombardment of nitrogen-14 atoms in the upper atmosphere by cosmic rays from the galaxy. Some of the carbon-14 atoms are incorporated into carbon dioxide molecules and these eventually find their way, via photosynthesis, into the tree rings. Strong solar activity deflects the cosmic rays; weak activity lets more of them through and more carbon-14 atoms are formed.
9. *Nature*, 1 June 1989.
10. Interviewed by the author, 24 August 1989.
11. See *New York Times*, 13 December 1989.

CHAPTER FOUR

The Global Commons: the atmosphere

The term 'rights of common' summons up for an English person an image of common land. This was once the unploughed wastelands where the tenants of a manor were entitled to graze their cattle, gather wood, cut turf and catch fish. Large tracts of them vanished in the agricultural enclosures of previous centuries and those that remain today are usually for leisure activities: walking and riding, bird-watching, flying model aircraft and the like. Even if it is often not strictly true – for many are privately owned – they are regarded as land held by the community for the benefit of the community.

The concept of the 'global commons' comes from the same stock. These commons are the oceans, Antarctica and the atmosphere. Place the history of the common lands beside them and the similarities can be seen. There is the general belief that they are there for all to use; and, equally, there are those who feel they have a right to restrict use and enclose parts for their own benefit. The lords of the manor in the case of the global commons are the rich, advanced nations with the ability to exploit the seabed and, potentially, mine in Antarctica. And the tenants are the developing nations who want a share in the wealth extracted from what they regard as common property. The idea was formalized in 1967 when Malta's ambassador to the UN, Dr Arvid Pardo, proposed that the deep seabed should be reserved for the benefit of all as 'the common heritage of mankind' whose mineral wealth would contribute to the development of poor nations.

On the fringes of the global commons are the great rainforests of the tropics, regions subject to national sovereignty, but nevertheless regarded by some as so vital to the equilibrium of the world climate that they are international rather than purely national resources. Beyond the atmosphere are the spatial commons of the moon, the space in the vicinity of the earth and perhaps the other plants of the solar system. And if all these are so vital and part of our common heritage, why is the Arctic not included? After all, it too is a polar zone essential to the maintenance of the earth's climate and large parts of it are uninhabited ice. The answer is that all the land on the rim of the Arctic basin is indisputably owned by someone and two of those someones are the United States and the Soviet Union. The Arctic is still, despite the ending of the cold war, a strategic zone separating the land masses of the superpowers who use its seabed rifts and ice to conceal their missile submarines. Unlike Antarctica, it is not neutral and even if it is neutralized before long, sovereignty over its shores is strongly established and its most accessible mineral resource, oil, is already being exploited.

Manipulating the weather

In the early 1960s the Americans in South Vietnam noticed that while the Buddhist monks rioting in Hué were undeterred by teargas, they always went home when it rained. The significance of this for riot control was quickly grasped. The problem was now to ensure that the rain arrived at the same time as the Buddhists. The CIA decided to experiment with a light aircraft which dropped silver iodide on the rain clouds at the appropriate time. It worked, and by 1967 weather manipulation had been given a new status as geophysical warfare. American aircraft operating under conditions of secrecy from Thailand regularly seeded the monsoon clouds over the Ho Chi Minh trail through North Vietnam and Laos.

The military were more enthusiastic than the State Department, which from the start was concerned about the wider environmental considerations and the international repercussions when news about military 'precipitation augmentation' leaked out, as it finally did in 1972.[1] The argument that it was more humane to drop rain than bombs did not prevail. Cloud seeding was stopped. Presidents Nixon's and Brezhnev's agreement at their 1974 summit that weather modification for military purposes should be banned led to the 1976 Convention on Prohibition of Military or Any Other Use of Environmental Techniques. By that time it was fairly clear that cloud seeding could deliver more rain over a given area than the clouds would normally unload, but that as a weapon it was of very marginal value. Its civilian use to improve rainfall, break droughts, disperse potentially devastating hail storms and provide water for hydro-electric power stations has had variable success, sometimes accompanied by claims from those who have suffered storm damage as a result. One day there will almost certainly be attempts to modify weather on the global rather than local scale with the aim of undoing damage and creating a more benign environment, but the knock-on effect is always a deterrent: what is an advantage in one country can be a disaster in another.

Nations continue to modify the atmosphere, not actively, for military purposes, but inadvertently or selfishly or for short-term but pressing economic reasons which create pollution. That is nothing new; burning has always been an easy way to clear land – slash-and-burn cultivation is as old as agriculture. Smoke from heath fires in south-western England is reputed to have damaged French vineyards in mediaeval times, and well before acid rain became part of the vocabulary, industrial fumes and smoke frequently had observable effects beyond the borders of the polluter. A Norwegian scientist speculated in 1881 on the linkage between British industry and the pollution reaching his country. But it was not until the 1930s that a landmark case in Canada proved that one country could take legal action against another to obtain damages for atmospheric pollution.

The Trail Smelter, in British Columbia, was built in 1896 to refine lead and zinc. It was never a healthy neighbour for the farmers of Washington State on the other side of the border, but in the second half of the 1920s the Canadian owners made its offence worse by building two 400-feet stacks and spreading the pollution further afield. By 1930 the stacks were pouring out 300–350 tons of sulphur dioxide daily and causing considerable damage. The body used to initiate arbitration was the US–Canadian International Joint Commission, established under the Boundary Waters Treaty of 1909. The treaty was intended to ensure that the rivers and lakes on the border 'shall not pollute on either side to the injury of health or property on the other'. An arbitration tribunal consisting of an American, a Canadian and a Belgian chairman was set up and eventually awarded the United States damages of $350,000. The legal principles involved were recognized by the Stockholm Conference on the Human Environment in 1972.

Pollution of the troposphere may not be new, but its magnitude in the second half of the twentieth century certainly is. Sir Crispin Tickell, Britain's UN Ambassador and author of a pioneering book on the international aspects of climate change,[2] did not exaggerate when he described the problems in a statement to the UN's Economic and Social Council in May 1989 as 'of a kind time-bound governments have never before had to deal with. The atmosphere knows no boundaries and the winds carry no passports.' It was in the industrial setting of Manchester that the link between pollution and acid rain was first observed in 1852 and it is the industrial sources which cause so much ill-feeling and concern. Sulphur dioxide leaves the smokestacks as a gas and is converted in the atmosphere to sulphuric acid. The droplets it forms are then dissolved into rain clouds. Nitrogen oxide – which comes chiefly from road vehicles – also has to form an acid (nitric acid) before it can be dropped in rain. Acid rain's effects are complex and, in the case of river and lakes, indirect: aluminium in freshwater becomes a lethal poison for fish, most dangerous when the level of acidity is only just within the limits of what is officially rated as acid rain. Aluminium is the villain once again in the case of trees. Acid rain mobilizes it and other heavy metals in the soil to toxic effect, while at the same time it weakens the trees by leaching away nutrients. One reason why northern countries are particularly affected is the capacity of snow to store pollutants, which are sometimes present in such heavy concentration that the snow turns black. When the snow thaws, a spate of highly acidic meltwater flushes into the streams and lakes, sending fish-kills soaring.

Acid rain has become a shorthand for all airborne pollutants. But there is also ozone – beneficial in the stratosphere, a pollutant in the troposphere – and ammonia (a contributor to acidification by complex chemical routes which does not travel more than about 60 or 70 miles from the packed cattle and pig pens where it is produced) and hydrocarbons and trace metals. Human lungs and kidneys suffer as well as the land and the lakes. *Waldsterben*, or forest death, has devastated forests, most notably in West Germany, Czechoslovakia, Poland and Switzerland, like an arboreal black death. More often, though, it is damage to trees, rather than their death, which is noticeable, with conifers losing up to 20 per cent of their needles. Drought, which might become more common under the greenhouse effect in Central Europe and North America, weakens the trees and worsens the effects of pollution.

Two of the countries most sensitive to the effects of acid rain are Norway and Sweden, both notable for their forests and lakes. Unlike many other victims, they are more sinned against than sinners, since their outputs of pollutants are comparatively small; most of the acid rain falling on them is brought by the northwards curl of winds which have passed over Europe's main industrial areas. Their situation was worsened in the 1960s when higher stacks built to dissipate pollution and reduce the volume falling on industrial districts led to acid rain being spread over much wider areas. Ninety-two per cent of the unfortunate Norwegians' deposits come from elsewhere – largely from Britain, but also, they suspect, from Canada and the United States. Norway's Forest Research Institute estimated in 1989 that half the country's forests had been damaged, with more than one-fifth in the 'moderate to severe' categories, while Sweden's toll included 4500 lakes without fish life. Its effects can be seen on Northern Europe's heathlands, too. Where there was once unbroken heather there are now large patches and seams of coarse grass, the result of the acidification which has taken place in the relatively short space of 30–60 years. This process of soil acidification can go down to depths of over a metre and poses a threat to underground water as well as to the fertility of the soil in the long term (although the evidence on this is still mixed). Acid rain is also a pernicious corrosive which erodes stonework and even railway lines and at its worst – as, for example, at Pitlochry, in Scotland, in 1974 – can be as acidic as vinegar or lemon juice.

The pollution which causes acid rain is of a different order from the greenhouse effect, although there are interactions. Nitrogen oxides are common to both and sulphur dioxide (the main component of acid rain) in clouds may lessen the greenhouse effect by deflecting solar heat that would otherwise fall on the earth's surface. Like the greenhouse gases, acid rain is a natural as well as industrial phenomenon.[3] In industrial regions, sulphur dioxide is overwhelmingly man-made, but globally, nature contributes 50 per cent through volcanic activity. Nitrogen oxides, too, come abundantly from natural sources, as scientists on board the British research ship *Charles Darwin* found during research in the north-west Indian Ocean at the end of the 1980s. Significant amounts are produced by bacteria in layers of oxygen-poor water around 800 metres deep. Biological production appears to be just as important, if not more so than man-made sources.

Cleaning up Europe's air

The Trail Smelter case may have broken new ground by declaring that air pollution was an international issue, but more than forty years elapsed before the Convention on Long-Range Transboundary Air Pollution was signed in 1979. The convention, the first of its kind, obliges the signatories (all of them European or North American) to 'limit and, as far as possible, gradually reduce and prevent air pollution including long-range transboundary air pollution'. There are clauses on consultations between victims and polluters, exchanges of data and information on air pollution control technology, and a requirement that any industrial development or change in national policy that is likely to create pollution will be notified. However, there is nothing on liability for damage and no binding provision on arbitration of a dispute. Polluter and polluted are left to work out their own solution.

The matrix of the treaty was the UN Economic Commission for Europe (ECE), created in 1947. The economic and social systems of East and West were diametrically opposed, and confrontation was almost a reflex action in most gatherings where the opposing blocs met, but there was limited room for consultations on the more technical aspects of matters such as trade relations and inland transport. The ECE was conceived as a cold war bridging organization which covered considerably more than Europe, stretching from the United States and Canada to the Euro-Asian massiveness of the Soviet Union. However, an apolitical UN organization of that kind was incapable on its own of moving very far down the road towards binding agreements. Co-operation in Europe could not exist without security in Europe, and the two were not brought together in one forum until the birth in Helsinki in 1975 of the 35-member Conference on Security and Cooperation in Europe. CSCE's membership includes honorary 'states' like the Vatican, but otherwise it is the same as the ECE's. It was Brezhnev's desire for a détente with agreements that would ease the Soviet Union's economic and political isolation for as low a price as possible which produced CSCE. Its mandate was essentially, despite the presence of neutral and non-aligned states, a continuing East–West dialogue based on the Helsinki Final Act with its three 'baskets' of issues: security, economic and technological relations, and human rights. Prominent among the items in the second basket was the environment, with control of air pollution heading the lists of fields of co-operation.

Brezhnev wanted the Soviet Union to be seen as a benign and reliable partner within the specifically European family, an enduring Soviet aim. The idea of 'all-European congresses' on the environment, transport and energy was floated by him at the end of 1975 and then placed by the Soviet Union on the agenda of the first post-Helsinki session of the ECE, in April 1976. Almost a year passed before the Norwegians took matters further and proposed an international convention on air pollution. And so, by processes slowed by the deterioration of East–West relations in the second half of the 1970s, the

Convention was agreed in Geneva in November 1979 and, after another three-and-a-half years, brought into force with its own secretariat in the ECE in 1983.[4] Since Helsinki, the ECE has involved itself in transboundary water pollution and hazardous chemicals in addition to air pollution and its environmental department is now the largest in its Geneva offices.

The transboundary convention had its political significance, but it contained no commitments to specific cuts in emissions. It was not until 1987 that there were enough signatures to bring into force a 1985 protocol formalizing the aims of Ottawa's '30 per cent club', whose target was a cut in sulphur dioxide emissions of that amount by 1993. Scrubbing out sulphur dioxide by fitting flue gas desulphurization (FGD) equipment is expensive, and the biggest polluters have tended to prevaricate and demand more research before accepting an obligation to take action. Some of those who signed up for the 'club' were suspected by those who did not of political cynicism, since there was no evidence that they had the resources, or the will, to achieve the target. 'It was nonsense – political chicanery,' said a British official, dismissing the '30 per cent club' in a conversation with the author. 'Targets have to be fair and realistic.' The recent revelations about pollution in signatory countries such as East Germany, Czechoslovakia and the Soviet Union seem to bear out that assessment.

A number of countries, including the United States, Britain, Poland, Spain and Yugoslavia, did not sign, but West Germany, a major smokestack polluter, did, fitting FGD equipment at great cost to its main fossil-fuel power stations. For a signatory like France, which by 1985 was steadily increasing the amount of electricity produced by nuclear energy, signing was no hardship. Canada, the recipient of large amounts of US pollution, also signed. The non-signing Americans have subsequently declined to commit themselves to a bilateral accord with the Canadians. 'Great progress' would be made, said President Bush after he discussed the subject with Brian Mulroney, the Canadian Prime Minister, in May 1988 . . . and left it at that. It was cost rather than indifference or lethargy which underlay the reluctance to take action, as could be seen in the following year when Bush threatened to veto the Senate's Clean Air Bill, on the grounds that it would be damaging to the economy. The Senate, which had been deadlocked on the Bill for ten years, eventually approved one in April 1990 which concentrated on power generation, vehicles and toxic chemicals, and, if it became law, would double the costs of pollution control to $64 billion a year. The more modest proposals of the Administration's revised version would, it was estimated, add $19 billion a year to industry's costs by 2005.

Britain, the other major non-signatory of the transboundary convention, took a more co-operative attitude towards a similar EC agreement, the Large Combustion Plants Directive, in which the targets were seen, eventually, as 'fair and realistic'. In September 1989 Britain became the first country to take steps to meet the Directive's goals of cutting sulphur dioxide emissions by 60 per cent by 2003 and nitrogen oxide by 30 per cent by 1998, based on 1980 levels. A protocol on nitrogen oxides (signed by Britain) was approved in November 1988.

Negotiation of the directive had not been easy – it took five years, with twice-weekly sessions the rule for much of the time. British doubts about the costs of compliance returned as the privatization of the electricity industry – which had already run into trouble over its nuclear sector – drew near. By early 1990 it was clear that the government had decided that the nearly £2 billion cost of FGD was so heavy that it would adversely affect privatization. The programme was to be cut back and sulphur emissions reduced by using imported low-sulphur coal and modern gas-burning plants. There was independent evidence that the directive could still be met, but there was also a feeling that the government was retreating from a more stringent approach to the problem of acid rain. Britain, which led the industrial world with the Clean Air Acts of 1956 and 1968 after a particularly foul 1952 fog killed 4000 people in London, is still the biggest gross exporter of sulphur dioxide in Western Europe, exchanging deposits with industrial regions as far away as Poland.

Pollution in the lower layer of the atmosphere has in some ways provided a better test-bed for a major international negotiation on the atmosphere than the ozone layer. Damage to the ozone layer is caused by a small range of chemicals, mainly CFCs, used in the manufacture of a limited range of products, which can be replaced by less harmful ones. Acid rain pollution, on the other hand, comes from power stations, factories and vehicles, from activities vital to the economic well-being of a modern industrial state. It is also less 'abstract' than depletion of the ozone layer; it can be experienced more directly, even smelt.

Apart from the fact that it created a framework for action on dealing with international pollution, there are two points worth noting about the transboundary convention. The first is that the problems which prevented it from being conceived and agreed much earlier were political, rather than economic or environmental. It was the breakthrough represented by the Helsinki Final Act which made it possible. It is a point which has to be kept in mind when considering the preparations for a climate convention. Even when needs are recognized, the political problems can still be immense. The second is that it is a product of conference diplomacy, the twentieth-century's contribution to the polished art which was once the preserve of urbane men of great discretion who rarely need to venture beyond the points of a neat triangle formed by chancelleries, foreign ministries and the more exclusive receptions.

Conference diplomacy is multilateral and all too often tediously open-ended. The delegates' function is ultimately to pave the way for, or arrive at, an agreement, but it is recognized that they may have to sit for years listening to one another's speeches before change in the international climate permits, at best an agreement, possibly just the solace of a communiqué, and at worst an adjournment. There is rarely a collapse, UNCLOS, the UN Conference on the Law of the Sea, went through three stages between 1958 to 1982 before arriving at agreements covering almost everything except one of the more important issues: the recovery of minerals from the deep beds of the oceans. The CSCE process is similar: its Madrid review conference lasted three years and its successor, in Vienna, began in November 1986 and closed in January 1989.

But to emphasize the tedium of conference diplomacy is in a way to miss the point. It includes every nation with a claim to relevancy, and even many who cannot claim it, and it is never exclusive. Formal diplomacy may shape the conference, but the subject matter is generally outside the province of formal diplomats, concerned as it often is with subjects like trade, transport and pollution. It is, in fact, a product of the complexity and variety of twentieth-century international affairs. Bilateral relations still reign supreme in nuclear arms control, but there are no atmospheric superpowers; we are all equals in the greenhouse and what we do or don't do to preserve the atmospheric commons will be decided in international conferences.

Notes

1. *New York Times* indexes, 1972 and 1974, and reports in *N.Y.T.*, 3 July 1972. See also Lynton Keith Caldwell, *International Environmental Policy* (Duke University, N. Carolina, 1984), p. 226.
2. Crispin Tickell, *Climatic Change and World Affairs* (Center for International Affairs, Harvard University, and University Press of America, revised edition, 1986).
3. For a useful guide to the subject of acid rain, see Don Hinrichsen in Edward Goldsmith and Nicholas Hildyard (eds), *Earth Report: monitoring the battle for our environment* (Mitchell Beazley, London, 1988). Also the paper by Göran Persson at the Toronto conference on airborne pollutants, June 1988; and Trevor Davies, *New Scientist*, 8 April 1989, on acidity in snowfalls.
4. For background to the Convention see Evgeny Chossudovsky, *'East–West' Diplomacy for Environment in the United Nations: the High-Level Meeting within the Framework of the ECE on the Protection of the Environment; a case study* (United Nations publication; sales no. E.88.XV.ST26. 1989).

CHAPTER FIVE

The Global Commons: the oceans

The first work of modern oceanography, *The Physical Geography of the Sea*, was published in the United States in 1855, the work of a Southern naval officer, Matthew Fontaine Maury, whose writing was generally considered better than his science. The rhythm of his prose and the contemporary thirst for popular science carried it through six editions in four years and ensured its translation into six languages. It began:

> There is a river in the ocean. In the severest droughts it never fails, and in the mightiest floods it never overflows. Its banks and its bottom are of cold water, while its current is of warm. The Gulf of Mexico is its fountain, and its mouth is in the Arctic Seas. It is the Gulf Stream.

It is an attractive idea to present the Gulf Stream as a great river, a Nile of the North Atlantic. Its tropical warmth has enabled the modern civilizations of north-western Europe to defy the rigour of their latitudes and flourish; and one of its sources, the North Equatorial Current, brought Columbus to the Americas. Maury's image is also a considerable understatement. The Gulf Stream is all the world's great rivers multiplied many times over, carrying 120 million cubic metres a second as it races along past Florida, but not much more than the Rhine to the Amazon when compared with the Antarctic Circumpolar Current. These tremendous currents are warped into the oceanic gyres of the northern and southern hemispheres, which plunging and surfacing over great distances and at great depths in obedience to changes in temperature and salinity, moderate the earth's temperature.

The oceans cover two-thirds of the earth's surface and their circulatory system shares almost equally with the atmosphere the task of transporting heat from the equator to the poles. They are slow conductors of heat, though, with time-lags of 30–40 years before they catch up with air temperatures, and the part played by them in creating climate is extremely complex and a long way from being understood. There is speculation that the unprecedented world temperatures recorded in the 1980s may be due to long-term fluctuations in ocean circulation. Large releases of carbon dioxide into the atmosphere from upwellings of deep water currents, sharp rises in temperature in the North Pacific and tropical Atlantic, the shifting of trade winds, changes in currents which are initially wind-produced, changes in climate as a result and the drying out of the grain belts of North America and Europe could be linked in a chain of events. The World Ocean Circulation Experiment which will be based in the new Centre for Deep Sea Oceanography in Southampton, England, will concentrate over the next 5–10 years on strengthening an area of knowledge which is widely regarded as the weakest in climate change research. But, 'never overflows'? There is no confidence on that score these days. The seas through which the Gulf Stream passes have risen almost imperceptibly since the nineteenth century, and elsewhere, too, on average by between 10 and 15 cm. The more alarming scenarios of only a year or two ago which had the sea rise rising a metre, plus or minus 50 cm, by 2010 have since been scaled down. This is largely because of a marked change in assessments of the effect global warming will have on Antarctica, where most of the world's fresh water is stored. While there will be melting around the edge, where the ice has been thinning in recent years, there will probably be more snow and ice over the continent. The net result will be that the icecap will grow, taking water *out* of the oceans (which hold more than 97 per cent of all the water on earth) rather than melting into them. A study of the Western Antarctic Ice Sheet by the University of Wisconsin's polar research centre estimated that its weight was likely to increase by between 100 and 400 billion tonnes a year.[1] The Antarctic and Greenland ice sheets are so cold, anyway, that, even under greenhouse conditions, it is difficult to imagine them melting appreciably in the foreseeable future (although any melting from Greenland is likely to affect North Atlantic climates by chilling the sea with meltwater). There will still be an increase in sea levels, but through thermal expansion of the oceans caused by the rise of world temperatures. Dr Pier Vellinga, The Director of the Dutch National Climate Change Programme, who heads the IPCC's sub-group on sea levels, believes[2] the rise could be between 30 and 50 cm by the middle of the next century and possibly a metre by the end of it. That sounds small, deep enough to paddle in but not swim, but such a rise in the context of global warming will probably be accompanied by higher waves and more frequent and more violent storms; and also it could be a step in a process, not the end of the process.

Hurricanes are created by high temperatures and they need a sea temperature of at least 27°C as a primer. Higher temperatures in the mid-latitudes would be unlikely to bring hurricanes to temperate north-west Europe and north-east America, but they could produce more and fiercer storms, like the October 1987 and January 1990 gales, which wreaked such devastation in southern England and parts of continental Europe. Even those storms were modest compared with Hurricane Gilbert whose winds reached more than 300 km/h as it tore across the Caribbean in 1988.

A hurricane, with its swirling vortex, is not, of course, just high winds, tremendous seas and a deluge. It brings with it a storm-surge – water sucked into the low pressure in its centre, the eye of the hurricane. The relatively modest surge which combined with a normal high tide to drown Canvey Island in Britain in 1953 was 2.4 metres; Hurricane Hugo which struck parts of the eastern Caribbean and South Carolina in September 1989 carried with it a storm-surge that reached 6.7 metres. Combined with a rise in sea levels, a storm of that order could deliver a mortal blow to island states like the Maldives in the Indian Ocean, where the highest point is only 2 metres above sea level, and some Pacific mini-states, notably Tuvalu and Kiribati, as well, of course, as that wealthy exception among island mini-states, Singapore. Even Venice, a city whose Plimsoll line is already frequently submerged, might be lost despite its flood barriers.

Endangered nations

It was the Maldives' President Maumoon Abdul Gayoom's pleas for help at the United Nations and the Commonwealth summit in 1987 which focused attention on the threat of extinction posed by higher sea levels to these small countries. Developed countries have the ability to protect their coastlines against rises of a metre or so over a century, but even 10 cm would be a disaster for the Maldives, where the international airport is only 50 cm above high tide and has already been flooded by the swell from a storm some distance away. The population has grown to 200,000 people of whom a quarter live in the capital Male, and the gentle beaches and the same treacherous seas have made tourism the main foreign exchange earner. Another former British colony, Guyana, a country nearly as large as Britain, faces the prospect that a rise of 50 cm would put the majority of its 750,000 population at risk. Some 90 per cent of its people live in the low-lying coastal strip protected by only a dilapidated seawall on which $260 million needs to be spent.

For reasons of poverty as well as geography, the larger countries most vulnerable to sea-level rises are in the Third World: Bangladesh, Egypt, some of the thickly populated coastal regions of India, parts of China, Thailand, Vietnam, Indonesia and no doubt many others. Bangladesh presents the worst problems. The size of England, it consists of low-lying rivers and deltas and its 110 million inhabitants probably suffer more natural disasters than any other comparable country. Cyclones batter them from the south and floods swollen by the run-off from the deforested mountainsides above the Brahmaputra's tributaries sweep into the country from the north, causing at least 400,000 deaths since 1960. In an estuary, where tides are constricted, a new high-water mark is not necessarily just a matter of adding the sea-level rise to the curve of the high tide. It could, in certain places, be 6 per cent more than that and 12 per cent when combined with a storm-surge.[3] That, of course, could apply anywhere, but it is an extra threat for Bangladesh, where half the land is less than 5 metres above sea level. To all this must be added the effect of subsidence caused by a drop in the water table as a result of extraction for agricultural and other uses.

The Woods Hole Oceanographic Institute used a scenario based on a maximum 79 cm sea-level rise by 2050 and 217 cm by 2100 for its study of the impacts on Bangladesh and Egypt.[4] On the assumption that there would be complete damming or diversion of the rivers draining into Bangladesh's delta, subsidence would add 65 cm to the effect of the sea-level rise in 2050, making a total real rise of 144 cm. It would mean the loss of 16 per cent of the habitable land, the displacement of 13 per cent of the population (which by then should be well over 250 million, if the statistical projections are borne out), and the loss of 10 per cent of GNP. With continued excessive groundwater extraction, subsidence might double to 130 cm, creating even more refugees and land losses. Using the same scenario, Egypt's problems look even worse. It has a population of 53 million, and, when the country's huge, waterless deserts are excluded, an area only slightly greater in size than Belgium and Luxembourg combined for its people to live, work and grow food on. A combination of sea-level rise and sinking land could mean, by 2050, the loss of 15 per cent of the habitable land, 14 per cent of the population displaced and 14 per cent sliced off the GNP.

Perhaps the sea-level rises will not be as abrupt as Woods Hole anticipated, but the question remains: what is to be done? Bangladesh has built hundreds of miles of embankments in its deltas and under its Action Plan (1990–5) will raise and strengthen embankments on the Brahmaputra, which will help control its endemic river floods. A French company has proposed building 4000 km of high embankments over twenty years at a cost of $10 billion. Others believe it would be more practical to build storm refuges and improve the storm warning system. The crude fact is that the cost of building extensive defences is several times greater than the economic value of what the defences would protect.

An interim report prepared in 1989 by Pier Vellinga's IPCC sub-group on sea-level rises estimated that a sea-level rise of one metre would affect up to 300 million people, but obviously it is hard to be confident about forecasts of that sort. Many countries had not replied by early 1990 to his requests for estimates of their costs, some, like India, because they did not have the facilities or people to make them.

Rich countries, on the other hand, can raise money to keep out the sea and the popular support for action will be strong, even if the costs are astronomical. The United States has some 12,000 miles of coastline, some of the most endangered areas of which are on the eastern seaboard. The Gulf of Mexico coast is low-lying and, in the case of the Mississippi Delta, already sinking at a rate which every year converts 100 km^2 of marshland into open water. The East Coast from New York to the end of the Florida Keys is virtually one long beach backed by lagoons. Even extensive dyking would not protect Miami's freshwater aquifer from salt water contamination. What storms can do even in present circumstances was demonstrated in 1989 by Hurricane Hugo, whose eye passed directly over Charleston. The cost damage was estimated at $1 billion. But while the destruction caused by storms continues to be costly, the toll of lives has been quite dramatically reduced by forecasting and education. In the United States, hurricanes killed 8100 people in the first decade of this century; in the 1970s deaths were down to 226. Similarly, Hurricane Gilbert's toll was reduced by accurate tracking and frequent radio reports. The World Meteorological Organization, which opened its International Decade for Natural Disaster Reduction in March 1990, noted in some introductory notes that deaths caused by cyclones had declined in Bangladesh recently. The reasons, it said, were better warnings, use of shelters, evacuation of people and the deployment of trained volunteers.

The Dutch, who build their dykes on the assumption that only once in 2000 years are they likely to be overwhelmed, have come up with a relatively modest figure of $5 billion for their improved defences. A one metre sea-level rise is manageable, in their view. Given a lead-time of twenty years for such extensive construction work, Britain will probably have to make some decisions within the next five to ten years. London and a number of other cities as well as valuable farmland are vulnerable. The North Atlantic is getting rougher and the waves more powerful, and the storms may get worse. Safety margins will have to be increased. The Institute of Terrestrial Ecology's study *Climatic Change, Rising Sea Level and the British Coast* (1989) put the cost of raising sea walls sufficiently to counter a sea-level rise of 1.65 m by 2050 at £2500–3000/metre, with a total capital cost of £5 billion; and possibly another £2 billion for replacing pumping stations and outfalls. If the more recent scientific calculations are accurate, the rise sounds like an overestimate. Nevertheless, no one is sure and even a more modest rise would have its dangers. But even at the levels used by the Institute, the cost (based on preliminary work by the Ministry of Agriculture and Fisheries) is not crippling: less than half the Trident missile programme (£11 billion) or the road-building programme for the 1990s (£12 billion).

The refugees

'We face the spectre of major human displacements,' declared Sir Crispin Tickell:

> It requires a leap of the imagination to work out the numbers on the move in the event of global warming on present estimates. A heavy concentration of people is at present in low lying coastal areas or along the world's great river systems. Nearly one-third of humanity lives within 60 kilometres of a coastline. A rise in mean sea-level of only 25 centimetres would have substantial effects.[5]

Sir Crispin, who delivered this speech at the National Environment Research Council Annual Lecture, made the leap and plucked down the figure of 300 million refugees (a low estimate, in his opinion) of the next century's 6 billion population. Whether by coincidence or not, it was the same as the figure arrived at in Vellinga's interim report for the number of people who would be affected (not necessarily as refugees) by a one-metre rise in sea-levels. How would we manage? Tickell wondered. Compassion might move us one way, our sense of self-preservation in another. Within a country the growing numbers of refugees would represent a dangerous element. Few countries outside the industrial world would be able to manage a continuing crisis. There would be a slide into the sort of chaos seen in Lebanon and Peru. Even in the developed countries, resistance to immigration had become popular politics. Sir Crispin could only suggest sustainable development, birth control, cutting carbon dioxide emissions through more efficient use of energy, more protection against rising sea levels, agreements restricting greenhouse gas emissions, and so on.

If you accept the definition given in the 1951 UN convention, a refugee is someone who has been forced by persecution to flee and seek sanctuary. The 12 million people cared for by the UN High Commissioner for Refugees (UNHCR) in 1989 were, officially at least, all in this category. UNHCR does not recognize the new populations of economic and environmental refugees as part of its remit, although *de facto*, and when mandated by the UN General Assembly, it can and does. The UN's position on environmental refugees is set out in the UN Environmental Programme's (UNEP's) booklet on the subject.[6] In the view of at least one senior UNHCR official, rewriting the convention in order to broaden the definition to accord with contemporary realities would be more trouble than it would be worth. Could it afford to do so, anyway? Jean-Pierre Hocké, the former UN High Commissioner for Refugees, noted in August 1989 that his office was short by $85 million of the $429 million programme approved by his executive committee for 1989. Contributions were at an all time high but so was the number of refugees. Their numbers had swollen by 1.5 million between 1986 and 1988 and UNHCR's annual expenditure had risen by 23 per cent.[7]

A political or economic refugee is dignified to some extent by the implication that a stand has been taken or a choice made in favour of initiative or betterment. Environmental refugees are usually victims pure and simple, though, of natural catastrophes and implacable developments such as the advance of deserts and, no doubt before long, inundations. In most cases, they are refugees within their own country who wait until they can return safely to their land or property. When Hurricane Allen hit the Caribbean and Mexico in 1980 it left 230,000 people homeless and 500,000 without the means of earning a living, but the make-shift structures of poverty are soon repaired. In Bangladesh the people who live on the coast co-exist with the dangers of floods: 'most of them do not migrate even when they know of opportunities elsewhere. The incentives for them to remain are strong and include traditions, family ties and the prospect of acquiring more land – as the demand for land in such disaster-prone areas would be low.'[8] Their loyalty (or short-sighted opportunism) is remarkable considering the casualties inflicted by floods. A. Wijkman and Lloyd Timberlake[9] calculate that 386,000 people were killed by the thirty-seven cyclones that struck Bangladesh between 1970 and 1981. That is only 100,000 fewer than the total number of fatal casualties suffered by the British Empire in the Second World War.

Such violent storms are not a new phenomenon. A storm-surge in 1737 drowned 300,000 people on the same coast and in 1876 a similar surge killed 215,000. Casualties may have declined quite sharply recently as a result of better preparedness, but if global warming brings more cyclones with higher storm-surges as well as higher sea levels, then Bangladeshi resilience will be tested by many more disasters before the population bows before a changed environment and joins the throng of refugees. Even at the current population figure, the Woods Hole estimates for loss of land and displaced people would mean 14 million refugees – as many as the number of all refugees in the care of UNHCR in 1989.

The scale of misery that emerges when the effects of climate change and population increase are brought together is so great that it is hard to imagine how natural catastrophes worse than any that have occurred so far can be averted in the next century. With hundreds of millions involved, it would be possible to do no more than help those on the fringes of the crisis. For the great majority, nothing, or very little, could be done, particularly if climate changes force the developed countries to spend more and more of their wealth and energies on ensuring their own survival. But that is not evident yet and no one has been able to undertake any credible costings. As things stand at the beginning of the 1990s, a sea-level rise of less than 50 cm within the next 40–60 years would mean that states like the Maldives would have to be written off as lost causes and their populations relocated. Those threatened states that could be saved would be given priority in defensive work. The economic value of what was being saved should not, on the whole, be a factor. The aim should be to protect the land. Programmes would be extended in the light of experience and the degree of the threat. Action of that kind would require a mobilization of resources similar to that which takes place as countries prepare for war. They have never, however, taken place a generation ahead of a war, and there lies the problem. Civilization's fate in the twenty-first century may depend on a very radical change in attitudes adopted in the last decade of the twentieth.

The Law of the Sea

The oceans may be passive interpreters of atmospheric changes, their levels as impervious as the rise and fall of the tides to direct human restraints, but what crosses, goes into and is taken from them are subject to a surprising number of agreements and conventions. They are both a lesson and a warning to those preparing comprehensive legislation on the oceans' partner in the global commons, the atmosphere.

UNEP's regional seas programme, inaugurated in 1974, covers ten seas (in some cases, parts of oceans) and involves 130 countries. According to UNEP, some 20 billion tons of waste are poured or dumped into the sea every year, most of them in coastal waters. Sewage is the major problem, with oil, chemicals and the detritus of food and drink processing not far behind. The first and most advanced programme adopted under UNEP's aegis was the Mediterranean's, approved at Barcelona in 1975. It has been estimated that a molecule of seawater circulates in the Mediterranean for eighty years before it is washed out through the Straits of Gibraltar; and presumably the same applies to a molecule of sewage or any other waste. The demands of the tourist industry as well as the growing disgust of many of the inhabitants of the 120 cities that border the Mediterranean prompted a clean-up which is expected to cost $10–15 billion between 1980 and the mid-1990s. At the other end of Europe the Baltic has its own clean-up agreement, as does the North Sea, which has a sea-bed oil industry to contend with in addition to the outpourings of the Rhine and several heavily populated countries. Algal blooms off the coast of Norway and Sweden fed by nutrients washed from Jutland farms and the deaths of thousands of seals (which may have had nothing to do with pollution) all fed the public pressures which no government can afford to ignore these days. The saga of the *Karin B* in 1988 and the pcb shipments from Canada to Britain in 1989[10] demonstrated that waste disposal is an international problem as well as a lucrative international business. Taking other people's wastes is regarded as demeaning and arouses strong national feelings. The dumping of toxic wastes in African countries led, in 1986, to the Organization of African Unity asking UNEP to draft a treaty controlling the export of wastes. It was approved in March 1989 in Basle but was immediately criticized as ill-prepared and weak as a result of having been hurried through. African countries attacked it for not offering them sufficient protection, but were unable to muster the cash to finance a special meeting on the measures needed to strengthen it.

Underlying a network of agreements covering everything from dumping oil at sea to polar bears and civil liability in the transport of nuclear materials is the Law of the Sea, a many-tentacled but still dormant octopus whose gestation began in 1958. It was finally delivered after twenty-four years, and at the third attempt, in December 1982 at Montego Bay, Jamaica. Although it was adopted with 117 votes in favour, it is still not in force. Signatures are not the same thing as ratification and up to April 1989 only forty countries had ratified, still twenty short of the number required to put it into force. Moreover, those who have not adopted the convention are not obliged to abide by it even when, and if, the required figure is reached.

The reason for the Law of the Sea's difficulties is to be found in a clash between the interests of the industrialized nations and what the Third World perceives as its rights. From the early 1950s, Third World countries, notably those in Latin America, pressed for broad economic zones covering sea-bed rights and fishing beyond their shores.[11] Countries like Peru, which stood to gain from control over the anchovy fisheries nourished by the upwelling of the chilly Humboldt current from the Antarctic, were the first to make claims. Iceland followed the same course in its three 'cod wars' with Britain, which it eventually won. Its 200-mile exclusive economic zone is now internationally accepted. To return to the metaphor provided by the English common lands, these are the enclosures. Beyond them lies the wilderness of the deep ocean bed, and that is where the clash occurred. The 1967 proposal by Malta's UN ambassador, Dr Pardo, that the deep sea-bed should be regarded as 'the common heritage of mankind' was enthusiastically adopted by Third World countries and is enshrined in the preamble to the convention, which states that:

> the area of the sea-bed and ocean floor and the subsoil thereof, beyond the limits of national jurisdiction, as well as its resources, are the common heritage of mankind, the exploration and exploitation of which shall be carried out for the benefits of mankind as a whole, irrespective of the geographical location of States.

The imperial nations that had exploited the Third World were not going to be allowed to loot the sea-bed bonanza of copper, nickel, manganese, cobalt and any other metals whose nodules were assumed to be liberally sprinkled on the ocean beds. Provision is made in the convention for the supervision of the new deep sea regime by an international authority consisting of an assembly, a council and a secretariat, with a special chamber of an international tribunal for the Law of the Sea, seated in Hamburg, to deal with legal disputes. The Authority would work through an Enterprise which would control work on the sea-bed in accordance with a 'formal written plan', setting limits and, through the Authority, reserving a proportion of the agreed output for itself; and it would transport, process and market minerals. Another provision guaranteed to set the advanced industrialized nations' teeth on edge stipulates that the Authority will initiate and promote the transfer of deep sea technology from developed to developing countries.

Only a few countries have the capacity to exploit the deep ocean bed, and the economic incentives to do so are not very great. They would be even less if the deep sea miners were compelled to conform to a licensing system, contribute part of the proceeds from mining in the interests of countries (many of them without access to the sea) which had invested nothing in the technology or the equipment, and in addition had to hand over the technological know-how. It was not surprising that the United States, Britain and West Germany, which possess the technology, have been opposed to the sea-bed regime. Some Western countries – France, for example – have signed the convention, but none has ratified it.

In the light of the experience with UNCLOS the approach to international legislation on the atmosphere has been wary. North and South will undoubtedly square up to one another on the question of who pays and the moral and historical debts owed by the developed to the underdeveloped. Transfers of resources and technology from rich to poor are bound to be an essential part of the negotiations, even if the issues are life and death and time short. The argument advanced in Britain and other Western countries for not acting unilaterally or as a club are that it is essential to get a global agreement and that can only be achieved when the scientific facts are presented and there is a higher degree of certainty about what the greenhouse effect will do to our lives. That seeming certainty may never be very certain given the complexity and randomness of the factors influencing climate, and the governments of most industrialized countries must know that by now from their own scientists. What they are waiting for is not 'certainty' of that sort, but guarantees – guarantees that they will not be commercially penalized by taking expensive clean-up measures ignored by their competitors; guarantees that collective action by the North (which in itself is not exactly indivisible given its own East–West divide) will not be nullified by a surge in carbon dioxide and other greenhouse gases from a South which feels it has a moral entitlement to pollute since history denied it an early start on the road to industrial wealth. The experiences of the Law of the Sea convention are by no means an exact model for an atmospheric convention, but they contain cautionary lessons.

Notes

1. *New Scientist*, 16 December 1989.
2. Interview with author, 25 January 1990.
3. Unpublished 1987 paper by R.A. Flather and H. Khandker, quoted in L.A. Boorman, J. D. Goss-Custard and S. McGrorty, *Climatic Change, Rising Sea Level and the British Coast* (Institute of Terrestrial Ecology research publication no. 1, HMSO 1989).
4. John D. Milliman *et al.*, *Environmental and Economic Impact of Rising Sea Level and Subsiding Deltas: The Nile and Bengal Examples* (Woods Hole Oceanographic Institute, Mass., unpublished), quoted in *State of the World 1989*.
5. Sir Crispin Tickell, British UN Ambassador, National Environment Research Council Annual Lecture, *Environmental Refugees: the human impact of global climate change*, 5 June 1989.
6. Essam El-Hinnawi, *Environmental Refugees* (UNEP, 1985).
7. Article by Jean-Pierre Hocké, in *Herald Tribune*, 22 August 1989.
8. El-Hinnawi, op. cit.
9. Anders Wijkman and Lloyd Timberlake, *Natural Disasters: Acts of God or Acts of Man?* (Earthscan, London, 1984).
10. The *Karin B* arrived in British waters in August 1988 carrying 2000 tonnes of chemical wastes which had originally been dumped in Nigeria by Italy. Britain, West Germany and Spain refused to allow the ship to unload, as did the Italian port of Ravenna. Eventually, the Italian government was obliged to order the Tuscan port of Livorno to accept the cargo.
11. See Bo Johnson Theutenberg, *The Evolution of the Law of the Sea* (Tycooly International Publishing, Dublin, 1984), Ch. 1; and Lynton K. Caldwell, *International Environmental Policy: Emergence and Dimensions* (Duke University Press, Durham, NC, 1984), pp. 107–10.

CHAPTER SIX

The Global Commons: Antarctica

Antarctica is almost as ownerless in terms of actual occupation as the depths of the oceans. Its ice sheet moves with the implacable slowness for which glaciers are famous. The average depth of the ice is 2450 metres (at its deepest, 4770 metres, about the height of Mont Blanc), and the weight is so enormous that the bedrock is in places depressed by 600–800 metres, so that much of it is below sea level. Yet Antarctica was the unlikely scene of the last imperial scramble. Seven nations staked their claims there in the first half of the twentieth century. As flag-planting exercises, the claims hardly rate as colonialism, since the continent is uninhabitable on any normally sustainable basis. Perhaps they are better described as pre-emptive territorialism, a process Britain started as early as 1908. Expeditions were sent, flags planted, sectoral lines drawn from the South Pole, but until the mid-1930s no one knew whether it was a continental mass which lay beneath the ice or a series of islands. By 1933 Britain and her dominions Australia and New Zealand had claimed about two-thirds of Antarctica, an area almost as large as Australia itself and part of a continent on whose mainland no man had stood until 1895 and whose largest native non-marine inhabitant is a 3 mm midge, *belgica antarctica*. For much of the world the claims were to have as little validity as, say, a claim by the American Apollo astronauts to the moon. It was like the sea bed, *res nullius*, territory outside the jurisdiction of nations.

Possession, not science, was the principal motive for

exploration until the 1950s. The military were often involved. Argentine and Chilean bases were manned by the military, British warships made frequent visits to the region, and even as late as the end of the 1960s the US presence there was overwhelmingly military.[1] The Americans mounted exercises involving nearly 5000 men in the late 1940s, and the first and so far only military confrontation occurred in 1952 at Hope Bay, in the Antarctic peninsula, when an Argentine detachment fired shots to drive off a British team sent to repair a burned-out base. The Argentinian, British and Chilean claims all overlapped, while the United States and the Soviet Union refused to recognize the claims made by any of the seven countries involved.

The 1982 Falklands War has been interpreted as being in part a struggle over the control of sub-Antarctic resources; and that war is one reason why the British Antarctic Survey (BAS) enjoys a favoured position in government funding among the thirteen research bodies within the National Environmental Research Council.[2] General Galtieri is remembered with affection in British Antarctica, if nowhere else. At the end of 1988 the British government decided to spend an additional £23 million over three years on research and improvements to bases. Ever conscious that the 1981 decision to scrap the patrol ship *Endurance* sent the wrong signal to the Argentines, Britain keeps it Antarctic image high. So do the Americans, with up to 1200 people housed over the year at the McMurdo base close to the edge of the Ross ice shelf. Even the Indians felt that national prestige demanded that the expedition in the first half of the 1980s with which they staked their claim to a voice in the continent's governance should be composed of military men, just to ensure nothing went wrong. Antarctica may never have seen a battle, but it has never been quite the science-dedicated continent of peace depicted by those with a foothold there.

Yet having established that competitive nationalism opened up Antarctica (as far as such a place can ever be opened up), it is not so easy to explain what they were competing for. The British Navy mounted 'Operation Tabarin' in the closing years of the Second World War to establish permanent bases and forestall the possibility that Germany would use its close links with Argentina to do the same thing. The Americans tried to keep the Soviet Union out of the Antarctic in the early post-war years, but it proved hard to build convincing geo-strategic arguments for exclusion around the possibility that the continent – only 2 per cent of which is ever free of snow and ice – held unimaginable lodes of titanium, uranium, chromium and a wide range of other desirable and not-so-desirable minerals; or to explain how an ice-bound Soviet base opposite the Cape of Good Hope (but almost 3000 miles away) could be turned into a tenable submarine hide-out and military airfield. Even those who held out the security of the Cape shipping route as a reason for supporting South Africa did not waste time on an assessment of the Soviet threat from Antarctica.[3] It was not like the Arctic, the main area for Soviet missile submarines, where ice provides cover and there are important targets well within range of missiles fired through 'polynyas' (the sometimes large breaks in the ice). Nevertheless, it was concern that the continent could conceivably be of strategic value one day that led to the 1959 Antarctic treaty's prohibition of any military bases, weapons or materials and the insertion of the signatories' right to inspect each other's bases.

If it is hard to make the strategic reasons sound convincing, that may be because they were never the real, the essentially human and bold twentieth-century reasons for going to Antarctica. The place was the ultimate challenge, a hyperbole of terrifying statistics, more remote, so far as most people were concerned, than the moon, whose features could at least be seen on most nights. The wind there never simply blew. It was a perpetual blast which could rage at 90 mph all day and at times reached over 130 mph, and which routinely required a man to have six-inch spikes in his boots just to stay upright. Temperatures in Antarctica range, season to season, from 10°C to 30°C colder than the Arctic, the record low (recorded in 1983) nearly −90°C. So cold, in fact, that in the interior – a plain as flat as Kansas – there is almost no precipitation and it rivals the Sahara as the world's largest desert. Captain Cook speculated on Antarctica's existence when he sailed deep into the southern oceans in the 1770s but did not sight it. The huge seas that build up unconstrained by land barriers repelled all but the most determined sailors. Ice was another obvious problem. During the bad ice years between 1888 and 1907 sailors reported that at times there was only 100 miles of sailing room south of the Horn;[4] and it was estimated that during the Antarctic winter from May to July 1905, of 130 ships which sailed from Europe for Pacific ports four were wrecked, twenty-two suffered severe damage and fifty-three had not arrived or were unaccounted for by the end of November.[5]

The first sighting of Antarctica was in about 1820. It was fur and whale oil that gave the region value then. The romanticization in the names of national and personal honour and even spirituality came later. 'I wanted to sink roots into some replenishing philosophy,' wrote Admiral Richard Byrd in an attempt to explain his rather unprofessional decision to man single-handed for the whole Antarctic winter of 1934 the Bolling advance weather base on the interior of the Ross ice shelf. 'I should be able to live exactly as I chose, obedient to no necessities but those imposed by wind and night and cold, and to no man's laws but my own.'[6] It was carbon monoxide poisoning, not cold, that nearly killed Byrd, but for him the wind had human properties of malice:

> There is something extravagantly insensate about the Antarctic blizzard at night. Its vindictiveness cannot be measured on an anemometer sheet. It is more than just wind; it is a solid wall of snow moving at gale force, pounding like surf [a reference to the suffocating drift contained in Antarctic blizzards]. The whole malevolent rush is concentrated upon you as upon a personal enemy. In the senseless explosion of sound you are reduced to a crawling thing on the margin of a disintegrating world; you can't see, you can't hear, you can hardly move. The lungs gasp after the air sucked out of them, and the brain is shaken. Nothing in the world will so quickly isolate a man.[7]

The last article written by the naturalist and painter Peter Scott (whose father, Robert, perished in the Antarctic in 1911 after reaching the South Pole) before he died in August 1989 extolled the Antarctic as 'the part of the earth that is least affected by humans'.[8] In all its huge expanse, there are only about 3000 scientists, none of whom is more than a temporary resident. But the issue which prompted Peter Scott's article was the exploitation of the Antarctic for its mineral wealth, a development which could one day end its isolation. 'We should have the wisdom to know when to leave a place alone,' he wrote.

A continent for science

Leafing through the meticulously detailed 'Chronological List of Antarctic Expeditions and Related Historical Events' it soon becomes evident that the world is *not* leaving the continent alone. Increasing numbers of intrepid yachtmen are sailing to its more accessible fringes; the number of tourists visiting Antarctica aboard cruise ships each year is now greater than the number of scientists working there. If yachtmen and tourists can reach Antarctica, why not oilmen and mining engineers? One day they may decide they have the technology to enable them to make money in Antarctica. And it is that prospect which has led to the argument over whether mineral exploitation should be licensed or banned completely from an Antarctic 'wilderness park'.

The forum for that argument is the Antarctic Treaty, the product of the successful scientific co-operation which developed during the 1957–8 International Geophysical Year (IGY). It was a time that saw the enshrinement of Antarctica as a 'continent for science'. The overall planning was the responsibility of the International Council of Scientific Unions (ICSU), which co-ordinated the work of 67 countries, 12 of them with bases (55 bases altogether) in Antarctica.[9] Britain, Australia and New Zealand by now favoured the idea of a neutral non-military Antarctic and a US note on 2 May 1958 proposing that the twelve 'base' countries should follow up the IGY with a treaty ensuring that Antarctica was used for peaceful purposes only was well received, even though the treaty meant a limitation of sovereignty for the seven parties (Argentina, Australia, Chile, France, New Zealand, Norway and the United Kingdom) with claims. (The other parties were Belgium, Japan, South Africa, the Soviet Union and the United States.) The treaty, which entered into force in June 1961, froze the situation: no one's claim was prejudiced; no one was obliged to recognize a claim; and there would be no new claims while the treaty (which is open-ended) was in force. In the 'interest of all mankind' Antarctica was to 'continue forever to be used exclusively for peaceful purposes and shall not become the scene or object of international discord'. That principle was maintained even during the Falklands War, when the Argentinians and British took part without rancour in meetings of the consultative parties. The treaty's emphases were on non-militarization, science and continued scientific co-operation of the unblemished sort demonstrated in the IGY. (In the Arctic, by contrast, scientific work is conducted on a contained, national basis.) There was no mention of minerals or mining.

Despite the dormant claims and the decision to limit the consultative parties to nations with the capacity to set up scientific bases, Antarctica, the 'continent for science', had become a recognized part of the global commons. The treaty was, after all, in the interest of the world community, not just some of its richer members. But it was not a UN treaty (although the Charter is mentioned) and there was in the new regime an ambiguity born out of its elitism. Keeping Antarctica safe for science and mankind meant keeping it exclusive and free from the sort of Third World versus rich world confrontation which has blighted the Law of the Sea. Antarctica, on the treaty's terms, was certainly a prize for science.

The global climate models constructed by computers are in agreement in predicting that the largest changes in surface temperature due to a doubling of carbon dioxide in the atmosphere will occur in Antarctica. The progress of climate warming (or the reverse) can be measured on the gauge provided by the depth of its ice-sheets and sea ice, and the deepest ice is a palaeo-environmental library containing the records of past temperatures and atmospheres. One can look back 200 years to the very beginnings of the industrial revolution and trace the steady rise of carbon dioxide in the trapped air bubbles; and then, one can put the results in the context of the analysis of the ice core drilled out by the Russians at Vostock and see how closely, for more than 150,000 years, temperatures and atmospheric carbon dioxide concentrations have paced one another.

Since the days a century ago when sailors complained of cramped sailing room between the Antarctic ice and the Horn, the ice has retreated. Even thirty years ago it would not have been possible for BAS, the British Antarctic Survey, to build an airstrip at its Rothera base, roughly a third of the way down the Antarctic peninsula, as then-and-now aerial photographs of the base show. When the airstrip is completed in 1990–1 it will be possible for BAS for the first time to fly directly between an Antarctic base and the Mount Pleasant airfield in the Falklands, a very considerable advance made possible by a milder climate. But a retreat of the ice is more of a threat than a blessing. It reduces the albedo (reflectivity) of the Antarctic region, making it more absorbent of solar heat. There is enough fresh water in the Antarctic ice sheet to raise global sea levels by 60 metres if it melts.[10] The West Antarctic ice-sheet, which includes the Ross and Ronne ice shelves, is regarded as of special interest because much of it is grounded below sea level. Its disappearance alone could raise sea levels globally by six metres. Much of the investigative work into whether the ice-sheet is growing or shrinking is focused on this area. Scientific opinion at the end of the 1980s favoured the view that the icecap would grow (regardless of any thinning of the marginal and sea ice) because of increased precipitation as a result of global warming. Answers may depend on monitoring by satellite including the use of satellite-borne altimeters to detect alterations in the height of the surface. Similar marine ice-sheets which existed in the northern hemisphere during past glaciations appear to have disintegrated rapidly, 'with the major changes taking place on a time scale of the order of 100 years'.[11]

The IGBP was launched in 1986 by the International Council of Scientific Unions and given a five-year period for planning before becoming operational in 1991. The Council, a non-governmental body composed of twenty scientific unions representing some seventy countries, is also the co-sponsor, with the World Meteorological Organization, of the World Climate Research Programme (WRCP). Both are studies of global change, with IGBP taking a larger time-frame – from decades to centuries – than WRCP, which has three frames, the smallest dealing with weather predictions in a span of one to two months and the largest concerning itself with climate variations over periods of several decades. IGBP's brief is 'to describe and understand the interactive physical, chemical, and biological processes that regulate the total Earth system', a daunting undertaking which will undoubtedly overlap with parts of the WCRP's work since virtually nothing in climate and global change is separable from anything else. Everything is 'interactive', particularly in the polar regions, where oceans, atmosphere and ice come together.

Antarctica is one of the main places where the ocean currents are 'ventilated': warm water upwelling to create polynyas and melt the undersides of the sea ice and the ice shelves; extremely cold, dense water containing a nutritious mix of nitrogen and phosphorous plunging down to form currents as deep as 5000 metres on the ocean beds, a stage in a circulatory system whose round-the-world cycle can take 100 years. The upwelling of deep water releases stored carbon dioxide into the atmosphere, probably more than the Southern Ocean's surface waters and plankton take out of it. The plankton blooms during the spring in the mix of waters of different temperatures, absorbing carbon and sinking with it when it dies, to be stored and carried away by the outgoing deep water currents. Warmer water as a result of climate change might lead to more plankton and hence the removal of more carbon dioxide. Another attribute of plankton is that it emits dimethyl sulphide which, taken up into the atmosphere, provides the essential particles around which moisture condenses and forms clouds. More clouds could offset the greenhouse effect. But the ozone hole, whose existence has been the biggest revelation by Antarctic science, lets in more ultraviolet radiation which can weaken and kill off the plankton, setting off a chain of developments which could harm the whole ecosystem.

The mineral mirage

Science, and the need to monitor global warming, have in the decades since the birth of the treaty overridden crude national ambition and possessiveness as the reason why the original twelve consultative parties (or most of them) are in Antarctica, but there is another reason for a broader interest, one which has existed since explorers first found coal and burned it to keep warm in 1907–9. Antarctica is Gondwanaland, a detached part of what was once a world-continent linking South America, South Africa and Australia. Its climate was once tropical: dinosaur bones as well as coal have been found there. If the countries it was attached to are rich in minerals, then, surely, so must be Antarctica; or so the argument goes.

When the Antarctic treaty parties held their 15th meeting in Paris in October 1989 there were 39 present, 22 of them consultative parties by virtue of having paid the entry fee with a scientific base. The three major Third World countries – Brazil, India and China – were among them. More Third World countries attended as signatories, the most recent being Colombia. It was the Third World which put Antarctica on the UN General Assembly's agenda repeatedly in the first half of the 1980s. Dr Mahathir, the Malaysian Prime Minister, was particularly active, with statements claiming the continent as a common heritage and dismissing the treaty as an agreement between a 'select group of countries' which did not reflect the 'just claims' of UN members. 'I have heard there is gold at the South Pole and I want a part of it,' he is reputed to have said, summing up much of the prevailing feeling. On King George V Island in the South Shetlands there are now nine bases. This is foothold territory, one of the more easily accessible parts of Antarctica during the summer, a base camp for the new Klondike when, and if, it is discovered. 'It's quite crazy,' a British scientist told the author. 'They have this notion that there will one day be a mineral bonanza and they want to be in there with a clean ticket. A lot of the work they do is minimal and often a duplication. As for the mineral wealth, it's a lot of cobblers.'

The Antarctic Treaty does not mention minerals, but like most treaties it allows for protocols and conventions. There are conventions on fauna and flora, seals and marine living resources, the last adopted in 1980. 'In 1959 . . . very few individuals could have foreseen that a decade-and-a-half later the Antarctic Treaty Nations would be faced with the possibility that there would be serious speculation on the exploration for and eventual exploitation of mineral resources, not only on the continent, but also on the floor of the southern oceans that surround it.'[12] What changed the situation was the development of an Arctic oil field on the Alaskan North Slope in response to rising prices following the 1974 oil crisis and a generalized concern in the first half of the 1970s about mineral reserves. If the technology existed to extract oil from the Arctic why should it not be applied to the Antarctic? The Nansen Foundation held a conference on Antarctic mineral resources in 1973; in 1975 the Antarctic twelve asked SCAR to assess the possible impact of mineral exploration and exploitation on the environment; and the first meeting on a mineral convention was held in the following year. There had already been rumblings that Japan, West Germany and France had been exploring for oil under the guise of seismic research of the sea bed. Concern that the 1980s would see a dangerous flare-up of disputes over Antarctica's presumed mineral wealth was heightened by the Falklands War:

> The 1982 war merely accentuated rather than caused Antarctica's enhanced position in the international arena, a position which derived essentially from a range of factors related to its perceived marine and mineral resource potential, changing political and legal ideas developed in the context of the New International Economic Order and the common heritage principle, the U.N. Law of the Sea discussions and environmental considerations.[13]

The Antarctic Treaty countries initialled their agreement to the Convention on the Regulation of Antarctic Mineral Resource Activities (CRAMRA) in Wellington, New Zealand, In June 1988, after a decade of negotiation. It established a mineral resources commission that would collect information on any proposed mining and then decide whether exploration should be allowed. The prospective rules would not allow deep dredging or drilling to depths of more than 25 metres into the rock. No development would be permitted unless it was deemed environmentally safe and it would be subject to licensing and monitoring by a regulatory committee. All seven countries claiming sovereignty over Antarctic territory had to sign by 25 November 1989 for it to enter into force. Five (Argentina, Chile, New Zealand, Norway and the United Kingdom) did so; Australia and France had second thoughts, even though they had taken part in drafting the convention and had initialled its terms. The Australian Treasurer, Paul Keating, was reported to have urged caution shortly after the Wellington agreement, but in terms very different from the objection registered later by his prime minister, Bob Hawke. Australia's sovereignty would be threatened by the convention, said Keating.[14] Australia would share a place on the commission with nineteen other nations and would not receive royalties on mining. In his view that meant conceding Australia's claim to 42 per cent of the continent for 'virtually nothing'.

The strongest supporters of the convention are Britain, New Zealand and the United States. They take the view that the convention is an important first step towards guaranteeing that Antarctica will be protected from uncontrolled exploitation. Tim Eggar, the British Foreign Office minister responsible for Antarctica, made what he agreed were 'apocalyptic assertions' about the need for the convention when he spoke during the debate on the second reading of the Antarctic Minerals Bill (which enacts the convention) in July 1989: 'There are issues at stake in the Bill that are critical for the future of the world environment. It gives us an important opportunity to protect our environment. If we do not take that opportunity, we risk inflicting catastrophic damage on our planet.'[15]

A World Park?

The danger was that at some future date exploitable resources would be discovered and there would be an attempt at unregulated extraction. In Australia, Dr Phillip Law, the grand old man of Australian Antarctic exploration, had been taking much the same view for some time. The idea of a permanent prohibition on exploiting Antarctic resources was 'naive idealism'.[16] But that was the idea adopted by both Hawke and the French prime minister, Michel Rocard, when the two met in Canberra in August 1989. They joined in calling for the creation of an Antarctic 'world park' administered under a new Comprehensive Environment Protection Convention. Both were under pressure from environmentalists in countries where electorates were turning increasingly green and they had already discussed ditching the convention when they met in Paris two months earlier. France had taken a public relations battering from Greenpeace at the beginning of the year over the building of an airstrip at the Dumont-d'Urville base, a development in which Adelie penguins were killed when their island nesting sites were flattened by blasting. Jacques Cousteau, the famous marine naturalist, was opposed to the mining convention and in favour of an Antarctic secretariat to protect the continent. Apart from whatever Rocard may have felt personally about the rights and wrongs of mining in Antarctica, a save Antarctica policy had the attraction that it might improve South Pacific relations damaged by the sinking of the *Rainbow Warrior* and nuclear tests at Mururoa Atoll. Hawke, for his part, is said to have been disgusted by Greenpeace's evidence of pollution at the bases, where disposing of waste is a continual and expensive problem. His opposition to mining in Antarctica was at least consistent with a decision to retreat from earlier approval to permit mining in the Kakadu National Park in Australia's Northern Territory.[17] A British diplomat had a different view of the reasons for Hawke's enthusiasm for the 'world park'. 'It was a cheap way of showing how green he was at a time when it was politically advisable to be green,' he said. As for the pollution problem: 'It's not much of a problem. Much of the criticism is about releasing human sewage into the sea, but it is no more biologically unacceptable than what comes from 15 million shitting seals.'

By the time the treaty countries met in Paris in October, France and Australia were not alone: Italy and Belgium had moved towards supporting their stand. Deadlocked on the mining convention, the treaty parties agreed to meet again in the autumn of 1990, in Chile, to discuss the protection of the Antarctic environment and a liability protocol as added protection against pollution. The pro-CRAMRA forces hoped the French and Australians would eventually come round. The idea of a permanent ban on mining was not, in their minds, a realistic political objective. National interests could eventually lead to its being ignored. 'I would predict that no one will go there for oil,' the leader of the British team, Dr John Heap, told the author. 'But it's not so unlikely that they won't·go there for hard-rock minerals [i.e. copper, chromium, etc.].'

But when the froth of argument about mining in Antarctica is blown away the exposed reality is as forbidding as ever: Antarctica is virtually unexploitable in the foreseeable future. The only metallic mineral found in quantities large enough to be rated as a deposit is iron ore in the Prince Charles Mountains, and the grade is so low that it would not be considered worth mining in Australia. The coal in the Transantarctic Mountains may have kept explorers warm, but it is of poor quality. And in any case, there is no shortage elsewhere of coal that can be mined at a fraction of the cost. Evidence is still short to support claims that the Dufek Massif contains an exploitable platinum deposit similar to the one in the Bushveldt Complex in South Africa. Platinum, copper, gold, silver, chromium, tin, nickel, cobalt, molybdenum, lead, zinc, manganese, titanium and uranium have so far been found only as traces, even in the relatively accessible Antarctic peninsula.

Oil may well exist on the continental shelf, but there has been no systematic exploration of the sort associated with sea-bed oilfields like the North Sea and the Alaskan North Slope. The Antarctic continental shelf is narrow and deep – 500–1000 metres deep and not at all like the shallow Alaskan shelf. Icebergs would present an enormous problem. Their draught – up to 1000–1300 feet – makes them subject to the pull of deep currents far below the surface. They scour the sea-bed, and, potentially, any equipment on it. It might be possible to tow away the smaller ones threatening rigs, but it is difficult to see how that would be done with icebergs sometimes as large as the island of Martinique. Even if it did prove possible to extract Antarctic oil safely, there would still be the problem of shipping it. Submarine tankers which could sail below the ice have been suggested, but no one has tried to cost an enterprise of that sort. Surface tankers would have to be very large and sturdily built to barge their way through the ice. The appalling oil spillage at Valdez in Alaska in 1989 is unlikely to dispose either oil companies or public opinion towards the use of that form of transport in an ecologically fragile area.[18]

Given the remoteness and intractability of Antarctica, the idea of a world park, with its instant image of an ice-bound Yosemite, seems somewhat fantastical. Even if mass tourism on land became feasible, hotels and camp sites would hardly be encouraged in a terrain where a footprint on a rare patch of moss can be seen for decades. Would the new regime be more conducive to scientific investigation and keeping Antarctica pristine than the present one? Probably not. For all the accusations that the treaty has placed Antarctica in the hands of a club of greedy rich nations, it has been successful in opening the continent to science and preventing it from becoming a preserve of the military or a dumping ground for nuclear wastes. The Antarctic Treaty works and at the same time provides continuous instruction on the complexity of green diplomacy – and on the problems the simmering conflict between the have and have-not nations will bring to the approaching negotiations on a climate convention.

Notes

1. Much of the information in these first paragraphs has been gleaned from Peter J. Beck, *The International Politics of Antarctica* (Croom Helm, London, 1986) and Robert Headland, *Chronological List of Antarctic Expeditions and Related Historical Events* (Cambridge University Press, Cambridge, 1989), both definitive works.
2. See NERC Chairman Hugh Fish's report (31 April 1988) in which he noted that, apart from Antarctica, total income for research in 1988–9 would be 4 per cent less in real terms than in the previous year.

3. Patrick Wall (ed.), *The Southern Oceans and the Security of the Free World: New Studies in Global Strategy* (Stacey International, London, 1977).
4. S. Orvig (ed.), *Climates of the Polar Regions*, Vol. 14: *World Survey of Climatology* (Elsevier, Amsterdam, 1970), p. 312.
5. *The Role of Antarctica in Southern Hemisphere Weather*, conference report, May 1981, University of Melbourne.
6. Richard E. Byrd, *Alone* (London, 1938).
7. Ibid., p. 150.
8. *TV Times*, 9–15 September 1989.
9. See Beck, op. cit., for a useful account of the IGY and the Antarctic treaty.
10. Document prepared by the Cambridge-based SCAR, the Scientific Committee on Antarctic Research in connection with the International Geosphere-Biosphere Programme (IGBP).
11. *The Role of Antarctica in Global Change: Scientific Priorities for the International Geosphere-Biosphere Programme (IGBP)* (ICSU/SCAR, Cambridge, 1989).
12. J.H. Zumberge (ed.), *Possible Environmental Effects of Mineral Exploration and Exploitation in Antarctica* (SCAR, Cambridge, 1979).
13. Op. cit., see ch. 1.
14. See *Nature*, 1 December 1988.
15. *Hansard*, 4 July 1989.
16. Paper read to ANZAAS conference, Canberra, May 1984.
17. See *New Scientist*, 14 October 1989 for accounts of both this decision and the line-up on the mining convention.
18. For a more detailed account of the hazards and complications of Antarctic mineral extraction see M.W. Holdgate and Jon Tinker, *Oil and Other Minerals in the Antarctic: the environmental implications of possible mineral exploration or exploitation in Antarctica*, Report of Rockefeller Foundation workshop at Bellagio, Italy, March 1979 (SCAR, Cambridge, 1979).

CHAPTER SEVEN

The Global Commons: the tropical forests

Almost any stately or wealthy home in Europe bears witness to the fact that the trade in tropical hardwood is not new. Thomas Chippendale's favoured raw material, mahogany, came from the forests of Central America. The frequency of *embarcaderos* and 'haulovers' – names associated with logging – on the maps of the Caribbean basin testify to what a flourishing business there once was in mahogany and dye-woods; the beaches near the mouths of many rivers are littered with antique trunks and ends that were rejected or lost. So, was Chippendale an early accomplice in the destruction of the rainforests and the intensification of the greenhouse effect? The answer is no. Selective logging in itself does not destroy forests and, unless the owners of the master's furniture decide to burn their possessions, they will not increase the amount of carbon dioxide in the atmosphere. Nevertheless, Chippendale would find his customers divided on the subject if he were alive today. The supporters of sustainable, well-managed forestry which permits regeneration would be on his side. The anti-loggers, who believe that sustainable forestry exists as little more than a propagandist phrase, would be working hard to cut off his supplies, since their case is that exports of tropical hardwoods encourage the destruction of the rainforests.

These forests, which less than a century ago were regarded as virtually limitless, are being lost through burning and logging at a rate which pessimists believe could bring about their total disappearance before the end of the next century.

(Optimists believe that only half would vanish by then.) Burning contributes about 16 to 17 per cent of the carbon dioxide which goes into the greenhouse effect every year. But that is only part of the argument. There is the biological diversity which such a hothouse spawns. It has been estimated that half the world's species of plants, animals and micro-organisms live in tropical forests, and cutting and burning diminishes their habitat and therefore their variety and prospects of evolution. As there are possibly 30 million species of which only a relatively small number have been described, the loss to science, particularly medical science, which has obtained many valuable drugs from the forests, is potentially immense. Many may already have been lost. There is, too, the question of the eviction of indigenous people and their rights. Harder to define and justify, but nevertheless valid, is the sense of aesthetic deprivation caused by the loss of the forests, a feeling that a primal stronghold of innocence has been violated and we are moving closer to a world in which human fertility and greed ensure that the only places where nature is not in chains are the deserts. The symbolism of the rainforests is one good reason why they figure so prominently in the campaigns waged by the green pressure groups.

Foresters divide forests into two main categories: closed – an unbroken canopy; and open – more than 10 per cent of the ground covered by trees, with grass and shrubs covering the rest. More than half the world's closed forests are in the developed, industrialized nations, countries like the Soviet Union, Canada, the United States and Finland. Sustainable forestry is viable in the developed world. Its forests have expanded in the twentieth century. France, for example, has a greater extent of forest land today than it had in the sixteenth century. Both types of forest exist in tropical countries, with 'closed' usually meaning rainforest. Most of the tropical closed forest is in the Americas and most of the open in Africa (and most of the shrubby bush country as well). It is the closed forest which attracts international attention, principally be-cause it is being cleared at twice the rate of open forest. Its total area is the equivalent of two-thirds of South America,[1] with rather more than 25 per cent in Brazil, 9 per cent in

Zaire, 6 per cent in Indonesia, approximately 3 per cent each in Peru, Angola, Bolivia and Indonesia, and the rest shared among some seventy other tropical countries. Nine of them are expected to lose all their closed forest by the end of the twentieth century, by which time only ten countries are likely to possess timber worth exporting. Compiling reliable estimates of how much forest is being lost is extremely difficult and involves satellite pictures, reports from government environmental organizations like IBAMA (Brazilian Environmental and Renewable Resources Institute) and 'ground truthing' (i.e. on-the-spot checks) by lobbying groups. Definitions are another problem. For instance, the Food and Agriculture Organization's (FAO) 1988 interim report on the state of forest resources in developing countries lists Somalia as being 99 per cent wooded, which to anyone who has been there would seem ludicrous; most of it is arid bush country.

The FAO and UNEP came to the conclusion in the 1980s that the annual loss of closed forest was running at 7.5 million hectares – twice the size of Belgium – to which can be added a loss of 3.8 million hectares of open tropical forest. UNEP described the destruction as 'less than had been feared' and noted that at this rate of decline it would take more than a century to halve the size of the rainforests. For comparison, the depletion is, at about 0.6 per cent, less, proportionately, than the decline in the acreage of agricultural land in Britain between 1976 and 1986, which averages 0.8 per cent a year.[2] Because it has such a large share of the tropical forests, Brazil has been the focus of most of the international protests against their destruction. Yet an international taskforce created by the World Resources Institute, the World Bank and the UN Development Programme described its rate of deforestation as 'relatively low', a rating that hardly seems to square with an FAO–UNEP estimate that 2.5 million hectares is being lost each year, and even less with WRI's revised estimate in mid-1990 which suggested the loss might be as high as 9 million hectares. In Zaire, the second largest closed forest country, the annual depletion rate is assessed as a relatively modest 0.2 per cent, a tribute not so much to conservation as to the difficulties of working there. There is only one Atlantic seaport; despite the Congo basin's many rivers, logs have to be hauled expensively overland to avoid rapids and other problems; and the security of the loggers in such an ill-governed country is poor. The backwardness of its national owner is, it would seem, the best guarantee of survival a rainforest can have. In the more accessible Ivory Coast – one of the richest and most developed black African countries – the depletion rate is a high 7 per cent.

By any account (and the estimates vary) very little tropical timber comes from managed, sustainable forest – only 0.125 per cent, according to a report for the International Tropical Timber Organization.[3] That might seem like an argument for banning imports, but would a ban save the forests? The effect of such a move would be marginal at best. Only 5 per cent of the timber logged for industrial uses in Brazil is exported; the rest is used internally.

Loggers, naturally enough, go to the places where it is easiest to work and make a profit. If there are wild trees near rivers on which they can float their trunks to a port, they will cut them in preference to less accessible trees or to putting money into slow-yielding plantations. But even going to the most accessible places usually means cutting roads and skid trails through the forest. A study in the Philippines[4] estimated that the tonnage of soil carried away from a logged rainforest with roads was 240 times as great as that from uncut forest. The roads were responsible for 84 per cent of the erosion. The silt carried down to the sea destroyed corals and killed fish. Erosion of that sort is not the end of the story. Logging roads open up the forest to the farmers. The FAO–UNEP assessment showed that 55 per cent of the logged forest eventually became deforested. However, logging is not the main reason for deforestation. Clearing and burning with the primary intention of preparing the land for agriculture and construction work is responsible for two-thirds of the overall loss.

The fuelwood gatherers

Logging's importance shrinks again when it is put in the context of the uses made of wood. FAO estimated[5] in 1985 that 80 per cent of all wood harvested in the developing countries was consumed as fuel – almost the exact reverse of the situation in developed countries, where 81 per cent was used for industrial purposes. The moral may be that those who wish to save the forest, covered or open, will have to think in terms of providing alternative cheap sources of energy – a huge and costly task.

Gathering firewood dominates the lives of millions of people. Where it is scarce or non-existent, as in cities, it is expensive:

> In some countries malnutrition is due not to lack of food but to the lack of fuelwood for cooking. Families are forced to eat less nutritious quick-cooking foods or even uncooked meals to an extent that impairs their health. Urban demand for fuelwood and charcoal is expanding the economic distance for clearing and hauling wood, leading to ever-widening circles of devastation around cities and towns. Without major policy changes to ensure better fuelwood conservation and increased supplies, by the year 2000 some 2.4 billion people (more than half the people in the developing countries) will face fuelwood shortages and will be caught in a destructive cycle of deforestation, fuelwood scarcity, poverty and malnutrition.[6]

The knock-on effect is considerable. A lack of fuelwood leads to burning dried cakes of cattle dung and the leftover straws and stems of crops. That, in turn, means less fertilizer – 400 million tons less manure in fuelwood-poor regions, according to FAO – reducing food yields by 14 million tons, nearly twice the amount of food aid delivered annually to poor countries. Fewer trees means more erosion. The run-off of rainfall increases and the water table drops. And that, in the drier tropical regions, translates into desertification. Nor is rainfall recycled locally through the transpiration of trees.

Above all, there is the immense amount of carbon dioxide released into the atmosphere by so much burning for one reason or another: possibly 20 per cent of that released by burning fossil fuels, and, at 16–17 per cent of the total, a major contributor to the greenhouse gases. Moreover, grass and farmland do not have the storage capacity for carbon that a forest has.

Because they have been exploited, used and subjected to the pressures of growing populations since the sixteenth-century colonization, the Central American forests provide a microcosmic preview of developments in the vaster expanses of Amazonia. The evidence can be seen of the regenerative powers of tropical forest, the industries it sustains, and its destruction. In Mexico's Chiapas province and its Guatemalan neighbour, the Petén, the temple pyramids which in places break the forest canopy were once the centres of extensive Maya communities which deforested the land for agriculture. The Maya went and the forests returned.

For centuries the forests have been cleared by slash-and-burn cultivation, leaving soil which usually produces reasonably good crops for two-and-a-half years, after which the weeds, new growth and declining fertility defeat the farmer. The forests have been largely cleared of their valuable timber in relatively recent times, and today the only Central American nations which export timber are Honduras and Panama. Even the huge cedars from which the long canoes are hewn are hard to find. But a walk through the forests reveals other industries: the criss-cross lacing on the trunks of sapodillas tapped for chicle, the raw material of chewing gum, and communities of rubber tappers gathered around cauldrons of boiling latex. Compared with the cathedral quiet of a northern forest, a tropical forest is a noisy, diverse place, rich in plant species and bird and insect life. A large tree can be an empire of its own, home to mosses, bromeliads, orchids, hummingbirds and monkeys. But the forest often ends before long, giving way to cattle ranches and the small, constantly encroaching farms which provide a meagre living as an alternative to existence in cities and rural areas where increasing populations demean and impoverish life. The only hope of saving the Central American rainforest lies in schemes like that for the two million hectare Maya Peace Park which would incorporate tracts of forest in Mexico, Guatemala and Belize.

The British planted rubber in Malaya and Ceylon and made

a fortune out of it. The Dutch planted teak in Indonesia and today all the teak exported from that country comes from plantations established eighty years ago. But in Central and South America it was always cheaper to harvest the wild forest. Before the days of the Asian plantations the Amazon boomed on wild rubber and Manaus, Amazonia's capital, built its famous opera house out of the proceeds. Few tropical hardwoods are planted commercially in the way that conifers are in northern latitudes. It takes a long time to get your money back on trees, and in poor countries few can afford to wait. Those that are planted are usually fast-growing eucalypts and tropical pines. Popular, exportable hardwoods, like greenheart and Brazilian mahogany, come from the wild. There is, of course, managed forest of the sort left as a legacy by some colonial administrations, notably in India, which has 32 million hectares of managed forest. Burma, Bangladesh, Pakistan and Malaysia have managed forests, too. But, generally speaking, the proportion of managed forest seems to be declining. The taskforce on tropical forests assembled by the World Resources Institute noted that 'One of the most difficult issues to address in planning forest development is this lack of commitment to forest management.' Even if there had been political commitment, there would still have been a lack of skilled personnel and technical knowledge.

The 'Year of the Forest' and after

1985 was 'The International Year of the Forest' and the World Resources Institute's taskforce worked closely with the FAO, which had produced its own five-point Action Plan focusing on fuelwood; forestry's role in land use; forest industrial development; conservation of tropical forest ecosystems; and strengthening forestry institutions in tropical countries, notably in research and training. WRI elaborated this framework to produce an $8 billion 'five-year (1987–91) program of accelerated action' with investment profiles for fifty-six countries. Its proposals included the incorporation of all rainforests into national land-use plans and their management to ensure their long-term sustainability. Pressure on the forests would be reduced by creating more plantations and intensifying agriculture on non-forest lands.

The Action Plan has been criticized on the grounds that it is little more than a catalogue of what should be done; the FAO is short-staffed and it takes too long to get projects moving. The cost of implementing the plan has been pared down to $5.3 billion and pledges by donor-nations were still short of that amount by the end of 1989. While it has won some credit for overcoming some of the difficulties caused initially by too little consultation with the forest countries and too much interference by the countries providing the funds, its friends among the campaign groups are few and far between. In an attempt to repair its damaged credibility, the FAO in mid-1990 proposed an international convention as part of a package of measures designed to remedy TFAP's failings.

Friends of the Earth, which has little time for the Action Plan, took an apocalyptic view of the situation in its June 1989 campaign literature on behalf of its Rainforest Appeal:

> Dear Friend,
> Imagine you are driving at midday. Your headlights are full on. People are wearing face masks. Others are choking on the acid fumes. Airports have been closed. Smoke hangs in an acid pall over half the country. That is what it was like last year in Brazil. 8,000 fires raged out of control across Amazonia on one day alone . . . .
> Friends of the Earth desperately needs your help now to prevent the greatest man-made ecological catastrophe yet known: the systematic burning of the Amazon rainforests.

An accompanying leaflet explained in more detail why money was needed. Two-thirds of Brazilian Amazonia had been earmarked for development and less than 4 per cent had been designated as a national park. Plans included the construction of seventy hydro-electric dams over the next twenty years, including one at Altamira which would have flooded 7000 square kilometres.

Such campaigning has been vigorous and successful. Visits

to Western countries by Indian chiefs, whose tribes were threatened by flooding and settlement provided emotive television footage. Sting, the pop musician, held a joint public relations exercise in the depths of the forests with Chief Paiakan, of the Kaiapo Indians, in protest against a dam. London had its first 'save the rainforest' ball in 1989. And the rainforests gained their first martyr in Francisco Mendes, a militant Brazilian rubber trapper killed trying to save his trees from the depredations of cattle ranchers. Public opinion exerted itself on governments striving to present green images and the pressure was transmitted to the World Bank, which, despite having funded a number of Amazonian hydro-electric schemes, withdrew backing for the big Altamira dam and some other schemes and announced that in future its loans would be directed towards 'environmentally protective' causes. In mid-1989 Brazil and Britain signed a memorandum of understanding on technical co-operation in saving the environment – the first of its kind for Brazil – and in November of the same year the British Prime Minister, Margaret Thatcher, committed £100 million over three years to the Tropical Forest Action Plan when she spoke to the UN General Assembly.

The year of the 'big burn', 1988, was not, in fact, nearly as bad as the preceding year, according to IBAMA. The 'burn' was down by about 40 per cent.[7] An exceptionally long and heavy rainy season combined with an anti-burning campaign by the Brazilian government is thought to have reduced burning still further in 1989. Pressure had worked – but at a high price in aroused nationalism. The Ecuadorian President, Rodrigo Borja, declared that neither his country nor the seven other members of the new Amazon Pact would permit external interference in Amazon affairs; and the Brazilian Army Minister, General Leonidas Pires Goncalves, attacked 'false ecologists' whose aim was to 'internationalize Amazonia'.

In fact, the repeated use of the name Amazonia and its identification with the national interests of eight countries had internationalized the region. Even though it was still a long way from becoming a recognized global commons, it had begun to sound like a regional commons.

The Brazilian government outlined its position in a

statement circulated in mid-August 1989 entitled 'Brazilian Policy on the Environment', which recognized that the Amazon rainforest was one of 'the last great ecological sanctuaries in the world'. IBAMA had adopted measures which ended all financial incentives for land clearance (tax subsidies for clearance are blamed for much of the destruction): all land development projects of over 1000 hectares would require approval by IBAMA; there was a new national forest fire prevention system; and a pledge had been given of greater protection of Indian lands, the demarcation of which would be completed by 1993. Then came the sting:

> To implement fully these ambitious proposals, however, Brazil will require not only the co-operation of foreign governments and non-governmental organizations, but also the technical expertise and financial support of multilateral institutions. This support could include the training of Brazilian personnel, free access to up-to-date technological and scientific know-how, as well as financial resources for the specific projects which Brazil is currently submitting to the World Bank and to governmental and non-governmental organizations in the US, UK, France and West Germany. The very scope and complexity of the measures being considered makes it difficult to put a precise figure at this early stage on the full amount of resources involved. However, the sum of $80 million that will be needed to put into effect an emergency fire prevention scheme in the region is an indication of the scale of support that is needed. Brazil also expects a contribution from those enterprises which already have a commercial stake in the development of the region.

The statement went on to note that other threats to the global environment originated mainly in the developed nations: 95 per cent of the CFCs and halons which depleted the ozone layer and three times as much carbon dioxide from the United States alone as was produced by Brazil's burnings, much of it savanna and cleared forests. Moreover, the issues of ecological protection in developing countries should not be considered in isolation from the wider context of their poverty. Brazil's hard-won trade surplus was largely being consumed by interest repayments on its foreign debt, rather than in overcoming the country's developmental problems. As a result, per capita income had been declining in recent years. The internal situation caused by that state of affairs risked being aggravated by 'international over-reaction to such a politically sensitive issue as the development of the Amazon region'. Objective long-term research into the rainforest was required if 'appropriate action is not to be frustrated by ill-informed debate and unfounded assertions', among them the claim that Amazonia is the 'lungs of the world' and deforestation affects the global climate.

Sustaining innocence

Rainforest accounts for about 40 per cent of Brazil, so it is not surprising that its government takes exception to its being regarded as a global commons which should be subjected to international control, whether through the carrot and stick of aid and World Bank funding or attempts to boycott its timber exports. It gained support for its stand at the May 1989 regional summit in Manaus of the Amazon Pact nations. They called for funds without strings, asserted their sovereignty over the region's resources and criticized the environmental damage caused by their creditors, the developed countries, to whom they owed a total of some $400 billion.

The cultural and economic divide along which developed and developing nations are ranged in the environmental argument can be seen at its simplest in the rainforests. The Western world sees the forests as a sanctuary which should, so far as possible, remain undefiled, and considers their indigenous peoples as the last innocents. Burning may give a climatic tinge to the argument by releasing carbon dioxide into the atmosphere, but essentially the case advanced by the lobbyists and their institutional and governmental allies is aesthetic and not so very different from the argument against mineral extraction in Antarctica. To the forest countries the whole thing carries a strong whiff of hypocrisy. Here are people who have grown rich by defiling their own environments protesting and even using sanctions to prevent poor people from making use of their natural resources. The Americans, they note, are unlikely to set an example by returning the prairies to the buffalo and the Indians. Nor are the Europeans about to restore the Alps to their pristine beauty by tearing down ski lifts and restricting tourism.

There are parallels in the course of events in the negotiations conducted under the Convention on International Trade in Endangered Species (Cites) on saving the African elephant, with ivory in the role of tropical hardwoods. In both cases the producers are in the tropics and the biggest consumer is Japan. Much of the argument revolved around whether ivory could be 'sustainably' harvested. The Europeans, Americans and East African countries such as Kenya (where tourism and game parks are a major industry) agreed that poaching was so well rewarded and intensive that only a global ban on the ivory trade would save the elephant from extinction. Zimbabwe, Botswana, Malawi, Mozambique and South Africa disagreed. They managed their elephant herds successfully and ivory was a valuable export. Japan, the market for 40 per cent of the ivory trade, sat on the fence during the negotiations, but faced with the prospect of a green backlash against its goods in many Western countries, eventually accepted a global ban accompanied by the creation of a panel of experts to decide if countries which kept the poachers at bay and managed their herds successfully could resume trading at some time in the future.

The nearest thing the rainforests have to a convention is the International Tropical Timber Agreement (ITTA), which was originally conceived as a commodities agreement similar to those which cover sugar, tin, rubber and so on. The complexity of the tropical timber trade (which deals in about 100 different species) and the steadily rising level of environmental concern in the 1980s saw it emerge as something rather different and unique: a commodity agreement between producer and consumer countries with strong environmental overtones. Conservation backed by sustainable forestry is recognized as a major objective. In 1986 ITTA begat ITTO (the International Tropical Timber Organization), whose agenda for its autumn 1989 meeting in Yokohama (where its secretariat is based) contained five projects concerned with sustainable management and conservation. Membership in 1989 comprises 23 consumer countries and 18 producers, with a total of 1000 votes on each side. National votes are based largely on size of production or consumption, with Brazil followed by Indonesia leading the producers, and Japan in the van of the consumers with 327 votes, followed a long way behind by the United States with 98 votes. The hybrid nature of the organization is reflected in the attendance at its twice-yearly meeting. The British delegation at the 1989 Yokohama meeting, for instance, was led by the Department of Trade and Industry but included a representative of the Overseas Development Administration. Organizations like Britain's Timber Trade Federation attend as advisers, while all the main green campaign organizations send teams to lobby. 'There has never been any channel before through which we can talk to producers,' said Geoffrey Pleydell of the Timber Trade Federation shortly before the 1989 Yokohama meeting. 'Now we have got it, I hope to God it works. It would be a great pity if it faltered.'

FAO, it is said, might not be unhappy if it faltered. ITTO's interest in the sustainable management of the forests duplicates parts of the Tropical Forests Action Plan's work. Timber trade organizations in consumer countries have their criticisms, too. They feel ITTO has been slow to put forward positive ideas and head off pressure for quotas or bans on imports. But ITTO is nevertheless a strong reason why none of the major green movements supports national bans on imports of tropical timber. No imports would mean no place for the banners on ITTO's council.

'If developed countries got out, ITTO would soon become an organization dominated by Japan and the producers,' said Francis Sullivan, the World Wide Fund for Nature's tropical forests officer.[8] Another expert recalled the Indonesian reaction when warned that public opinion might force European governments to ban tropical timber imports: 'What they said in effect was "Sod Europe! If you don't want our timber we can sell everything we've got to the Pacific rim countries." And they could, too. There is a terrific market there and its going to get bigger as China takes more.'

Early in 1989 the European Community's timber trade organization, the Union pour le Commerce des Bois Tropicaux, urged the EC Commission to adopt a two-pronged approach to conserving the rainforests. It should 'without delay' intensify efforts to ensure better forest management and at the same time approve a surcharge on all tropical timber imports. A similar idea had been taken up earlier by the British and Dutch timber trade organizations. Their idea was that the money raised would be channelled through a special ITTO fund to producer countries seeking help to improve their forestry methods. The British federation presented the government with a consultative document, but the reaction was non-committal. The government seemed unwilling to act unless there was general agreement on a surcharge within the EC. But more importantly, perhaps, it did not want to move ahead of a wide international agreement on environmental issues which would include aid, debt and training and other help in introducing sustainable development. Conservation groups and even individuals have been less coy about ransoming forests held hostage by threats of clearance. 'Debt-for-nature' swaps involve buying bad debts at a substantial discount from commercial banks who prefer getting some money in hard currency to no money at all from Third World debtor-nations. In return, the debtor 'pays' the new creditor by entering into agreements on land management and conservation. Swaps of that nature have taken place in Ecuador, Bolivia, Costa Rica and the Philippines and have received a qualified blessing from Mostafa Tolba, the UNEP Director. Not surprisingly, the 'debt-for-nature' concept has been fiercely attacked by Brazil and several other countries on the grounds that it undermines sovereignty and, by paying off small debts amid wide publicity, deflects attention from the problem presented by massive Third World indebtedness.

Tolba has described the problems of preserving the tropical forests as even more complex than those presented by the depletion of the ozone layer. He has suggested agreements involving conservation and trade regulations as well as a regional action plan for the Caribbean and Latin America which would generate finance internally as well as externally. But between words and plans and acts there are wide gaps and progress is hesitant. The lack of reliable information about the extent of destruction remains a formidable obstacle. Nationalism (and poverty) will have to be paid off before the tropical forests became accepted by their owners as global commons. The fact that Brazil will be the venue for the 1992 UN Conference on Environmental and Development offers hope. As host, it will be under pressure to be positive, which could open the way for a convention covering rainforests and the protection of the threatened bio-diversity which they above all places harbour. It should be easier, too, to deal with the subject within the context of a meeting which will aspire to stitching together the strands of a wide North–South environmental accord.

Notes

1. UNEP Environment Brief no. 3.
2. *Annual Abstract of Statistics* (Central Statistical Office, HMSO, 1988).
3. Duncan Poore, *Natural Forest Management for Sustained Timber Production* (ITTO, 1988).
4. Gregor Hodgson and John A. Dixon, 'Logging versus Fishers in the Philippines', *Ecologist*, vol. 19, no. 4, July/August 1989.
5. *Tropical Forest Action Plan* (FAO, Rome).
6. *Tropical Forests: A Call for Action*, Report of an international task force convened by the World Resources Institute, the World Bank and the U.N. Development Programme (Washington, DC, 1985), part 1, p. 6.
7. *The Independent*, 4 August 1989. The figures given were 74,444 square miles in 1987 and 46,706 square miles in 1988.
8. Interview, 18 October 1989.

CHAPTER EIGHT

The ozone layer

Even with the most burning issues, even when the smoke is visible, it can take time to convince governments of the nature and direction of the fire. In the case of the depletion of the ozone layer, it took time to convince scientific opinion, too. Here was a brand new issue. Scientists had predicted the greenhouse effect, but no one had foreseen that something as useful as a non-toxic and non-inflammable refrigerator coolant invented at the end of the 1920s could be indirectly lethal once it drifted into the stratosphere. There was some prompt but limited action in North America and Scandinavia a few years after the first warnings in the early 1970s, but wider international action had to wait. Diplomacy is a world in which running is not advisable. Agreements are reached only when, by processes as much psychological as rational, a subtle balance of anxieties, confidence and national interests is achieved. That applies to a major environmental issue as surely as it does to arms control. It is not necessarily a leisurely process, but inevitably it is a slow one.

The depletion of the ozone layer by chlorofluorocarbons (CFCs) was first recognized as a serious problem in 1974, but the scientists were not unanimous. In April 1976 UNEP's governing council decided to convene an international conference on the ozone layer. Nine years passed before the Vienna Convention for the Protection of the Ozone Layer was signed on 22 March 1985. The first protocol limiting consumption and production of CFCs and their close relations halons came into force on 1 January 1989. Some fifteen years had passed since they were first fingered as ozone depleters, in which time atmospheric concentrations of the most damaging CFCs had almost doubled, to the detriment of the world climate as well as the ozone layer, since they are a formidable greenhouse gas, too.

Nevertheless, the ozone layer treaty is on the whole a success story. It is the first treaty of its kind, a pointer, if not exactly a paved road, to the hoped-for agreements in the 1990s on carbon dioxide and other greenhouse gases. It is also a story with several morals, the first of which is always to look a chemical gift horse in the mouth. The gift in this case was the invention by Thomas Midgley, a Pennsylvania-born engineer, of CFCs as replacements for existing refrigerants such as ammonia. In a famous demonstration at a scientific gathering in 1930, Midgley proved the non-toxic qualities of his invention by breathing in the gas and then exhaling it to extinguish a lit candle. (Some years earlier Midgley, who has rightly been described as a maker of environmental time-bombs, had invented leaded petrol.)

A second moral is that introducing a new gas can be the atmospheric equivalent of importing cats into Australia; man's friend becomes the enemy of species which hitherto have lived in equilibrium with other species. CFCs are durable as well as predatory. The atmospheric lifetime of the most widely used one, CFC-12, which is used in foams, aerosols and refrigerators, is 139 years. Even the shortest-lived fire extinguisher halon, 1211, lasts for twelve years. That means that their drift upwards into the stratosphere, where the ozone layer is, can be very slow and they will still arrive intact. The evidence points to their not having begun to reach the atmosphere in significant amounts until after aerosol insecticide sprays were invented towards the end of the Second World War, to be followed within the next decade by aerosol cans for almost every conceivable domestic use. By the 1980s they had become a multi-billion dollar industry, with the United States leading consumption with an average of 1.3 kg per person in 1986.[1] At that time, about 96 per cent of the world production and 75 per cent of the consumption was in the industrialized nations. They had become a staple ingredient of a wide range of manufactures, including air conditioners, cleaning agents for high-tech equipment such as computers, blowing foams for building insulation, car seats and fast-food containers, as well, of course, as refrigerators and aerosols. It is reckoned that more than 25 per cent of the world's food supplies are refrigerated at some stage, so effective, cheap coolants are no mere luxury item, nor is it possible to introduce alternatives to CFCs without considerable financial pain. Du Pont, the American chemical firm, estimated[2] that restructuring its CFC business to produce ozone-friendly alternatives known as HCFCs (or HFAs) would cost more than $1 billion in the United States alone. Since the user industries would require similar expenditures to design and restructure their equipment, the costs to producers and users would be 'staggering' during the 1990s. Moreover, the most popular replacements, HFC 134a and HCFC 123, could, despite their quicker breakdown rates, still turn out to be potent greenhouse gases. The June 1990 London conference on updating the Montreal protocol agreed in principle they should be phased out by 2040.

Anxieties about CFCs had been stirred initially by warnings that emissions from a new breed of giant supersonic airliners known as SSTs might damage the ozone layer. The SSTs were scrapped for economic reasons, but two University of California scientists, Professor Sherwood Roland and Dr Mario Molina, soon discovered a real destroyer of ozone. They realized that all the CFC molecules that had ever been released were still active in the atmosphere. Their first calculations of what they were doing indicated a catastrophe was in the making. The findings were made public in 1974 and were found less than convincing by a number of scientists and, understandably, by the manufacturers of CFCs. The public in the United States, however, had heard and read enough to convince it that even if no one could offer direct proof of what was happening in the ozone layer, the use of CFCs in aerosol cans for frivolous purposes like hairsprays could not be condoned. Ozone, a thin layer in the stratosphere at the best of times, screens out most of the ultraviolet radiation reaching the earth's atmosphere. Depleting it would let UV-B through in greater quantities to cause more skin cancers and eye cataracts. it would damage more than human life; there would be serious harm to forests, crops and the plankton and shrimps which form the first line in the ocean's food chain. The Federal Government announced in the spring of 1977 that the use of CFCs in aerosols other than those for medical and other essential purposes would be phased out. Canada, Norway and Sweden followed suit. In 1980 the European Community agreed to freeze industry's capacity for producing the two most widely used CFCs, 11 and 12, and called for a 30 per cent cut in their use as spray propellants by 1982.

It looked like a good start, but that was not the way it turned out. For the EC to freeze its capacity was all very well, but, it was noted,[3] the capacity was far greater than actual production, which could thus continue to grow quite legitimately. The early gains from phasing out aerosols were soon lost as CFCs were increasingly used in foam blowing and refrigeration. By 1984 emissions were back to their 1977 levels and increasing by 5 per cent a year. Despite this loss of ground amid a lull in public concern, there was continued movement towards a treaty. A World Plan of Action on the Ozone Layer had been agreed in Washington in the spring of 1977 and UNEP was given the task of setting up a co-ordinating committee drawn from other UN bodies, the voluntary organizations and countries with a strong scientific base. By 1980 the evidence was mounting that the ozone layer was being seriously depleted and in the following year UNEP set up a working group to prepare a global framework convention. These conventions are interesting structures: they define the problem and the area of activity, establish general principles and obligations, funding arrangements, create a secretariat and set the goal. In this case the goal was to be defined in the eventual treaty as a protocol 'to control equitably global production, emissions and use of CFCs'. UNEP's models were conventions governing regional seas in which a number of countries had shared interests. The Helsinki accords which established areas in which the signatory nations were committed to working to improve East–West relations in Europe are a political version of the same concept, the framework in that case being of fairly palatial dimensions.

The hole in the layer

The 1985 Vienna Convention on the ozone layer is frequently cited as a model for a framework convention on climate. Above it would be built more impressive legal structures. UNEP's Deputy Executive Director William Mansfield suggested in August 1989 that side by side with the framework convention it would be necessary to construct a comprehensive Law of the Atmosphere, just, it seemed, for extra strength.[4] So much emphasis on legal edifices can be puzzling, particularly when one looks at the impressive dimensions of the Law of the Seas and remembers that it is still not in force. The steps towards controlling CFCs prompt questions about whether it was necessary to have both an ozone convention and a protocol to the convention, with two years between them. UNEP would have liked convention and protocol in one package, and that would seem sensible to most people. The answer is that the sense of urgency was not quite sufficient; nor was trust, for that matter. Principles had to be established first. Economic priorities had the whiphand over the environment. It took several more meetings before the protocol was signed at Montreal in September 1987.

What dramatized the situation more than anything else was the notorious hole (in reality, a thinning caused by depletion) in the ozone layer over the Antarctic discovered in 1982 by a British scientist, Dr Joe Farley, working at the British Antarctic Survey's Halley Bay research station. It was easier to understand than ozone molecules and it gripped the imagination. By 1984 it was clear that something quite unexpected and unprecedented was happening. Halley Bay had been taking regular measurements of the ozone layer since 1957 and there had often been substantial changes in its thickness, partly as a result of sunspot activity, but never anything quite dramatic as this. The American National Aeronautical and Space Administration's (NASA) Nimbus 7 satellite failed initially to confirm the findings because it had been programmed to discount anomalies as large as those reported by Halley Bay. Farman's analysis showed that each Antarctic spring between 1975 and 1984 had seen the layer thinned to almost half its usual thickness. His account of his finding arrived at *Nature*'s office in London at the end of 1984, but was not published until May,[5] too late to have any impact on the delegations negotiating the Vienna Convention, which was signed in March.

There were still at that time three basic theories[6] put forward to explain the springtime decrease in Antarctic ozone since the mid-1970s (prior to that time it had been slowly increasing since 1960); it was caused by CFCs and halons (from the same chemical family as CFCs and used in fire extinguishers); there had been changes in the circulation of the atmosphere which had begun to transport ozone-poor air into Antarctica; it was a cyclical destruction of ozone caused by periodic increases in the oxides of nitrogen produced by solar activity. Three American scientific agencies joined forces with the Chemical Manufacturers' Association to mount a field measurement campaign from the McMurdo base between August and November 1986. The findings pointed to CFCs and halons as the culprits but were not absolutely conclusive. An even bigger operation was mounted in the following year, this time with two aircraft – a DC-8 and an old U-2 spy aircraft – and 150 scientists based at Punta Arenas in the southern tip of Chile. Antarctica's circumpolar vortex dominates the stratosphere from April to October, creating the extreme cold in which stratosphere clouds composed of ice particles develop. These, it had been noticed, had become more persistent since 1984. The U-2 flew through the vortex and into the clouds. The readings it brought back proved that those who thought they played a key role in creating the ozone hole were right. Concentrations of chlorine monoxide in the clouds were 100–500 times greater than concentrations at the same altitudes in the mid-latitudes. From that it could be deduced that chlorine-containing chemicals such as CFC molecules stuck to the ice crystals in the clouds within the intense cold of the polar vortex. The reappearance of the sun at the end of the polar winter delivered the ultraviolet radiation which triggered the destruction of the ozone by releasing the chlorine atoms. Since one chlorine atom is capable of breaking up 100,000 ozone molecules before it fades away, the destruction was enormous.

In October, as the scientists left Punta Arenas, measurements showed that 1987 was by far the worst year on record for ozone depletion (it was matched in 1989 after a drop in 1988). In some places 95 per cent had been lost. As with most things to do with the atmosphere, the role of ozone is not confined to one thing. It is a minor greenhouse gas as well as a shield against ultraviolet radiation. Among the developments noted by the scientists in Antarctica was a decline in the temperature of the lower stratosphere in October and November. This suggested that the depletion of the ozone layer had a significant impact on the normal springtime warming of the stratosphere.

If the ozone hole was confined to a vortex spinning over the Antarctic ice sheet, it might be possible to regard it as an isolated phenomenon. But the Antarctic is what meteorologists know as a sink: heat is transported to it by the atmospheric circulation and radiated outwards. Ozone is brought to Antarctica by the same process. As ozone appears to have decreased by 5 per cent throughout each year since 1979 in all latitudes south of 60°S, there is fairly evidently a dilution over a wide area; even further than that, perhaps, for as the vortex broke up in the Antarctic summer of 1987, the December ozone levels dipped as far away as Melbourne in southern Australia. Fair-skinned Australians and New Zealanders are the most vulnerable people in the world to skin cancer caused by exposure, whether through sunbathing or outdoor work, to high doses of ultraviolet radiation. The incidence of malignant melanoma – the most virulent and deadly form of skin cancer – has grown by 5.5 per cent a year since 1982 in Victoria. South Australia and Queensland have seen their incidence of the same disease rise by 40 per cent between 1982 and 1987. UNEP's International Committee on the Effects of Ozone Depletion has suggested that the effects will be even more serious and widespread. The germination and flowering of many plants will be inhibited and crop yields will drop. Human immune systems will be less effective. But perhaps worst of all, the oceans' phytoplankton populations will decline, and that, apart from affecting the food chain, will mean a drop in the oceans' ability to absorb carbon dioxide from the atmosphere. 'A ten per cent decrease in carbon dioxide uptake by the oceans would leave the same amount of carbon dioxide in the atmosphere as is produced by fossil fuel burning,' the committee warned.[7]

The Montreal Protocol was agreed while the scientists were still in Punta Arenas, which meant that their promptly announced findings were confirmation rather than cause of the need for urgency. All the major producers were present in Montreal, as was a good cross-representation from the Third World. China, which had attended the Vienna Convention as an observer, was a participant. India, which had had no representation of any sort at Vienna, was present in Montreal as an observer. The third member of the developing world's big league, Brazil (whose per capita consumption of CFCs is around 100 g a year compared with the United States 1.3 kg) was a participant in both cases. It was a much better agreement than many had expected, setting firm limits with some elasticity for the developing countries. Consumption of five CFCs would be frozen at 1986 levels from 1 July 1989 and cuts would reduce consumption by 50 per cent by the year 2000. Consumption of halons would be frozen at 1986 levels from 1 February 1992. Production, on the other hand, was actually allowed to grow by 10–15 per cent above the 1986 level to allow exports to meet some of the growth in demand from developing countries. A ten-year grace period before implementation of the schedule of cuts was granted to developing countries whose annual per capita consumption was less than 300 g.

By the time the protocol came into force at the beginning of 1989 twenty-eight countries and the European Community had ratified it. But of those only five – Egypt, Kenya, Mexico, Nigeria and Uganda – were developing countries. Brazil, China and South Korea had been involved in the negotiations but had not signed. India was clearly not hurrying forward either. All were involved, on a small scale, in manufacturing CFCs and there was a suspicion that they might be ready to supply markets from which the signers of the Montreal Protocol would eventually be barred. Peter Usher, a senior official with UNEP's Global Environment Monitoring System, took the charitable view in an article written for a UNEP magazine[8] that the reason for their reluctance 'probably lies largely in their lack of awareness of its relevance to them. They are absorbed with what must seem like the more crucial matters of development.' Another UN official put a sharper and less diplomatic spin on the answer in a conversation with the author: 'They see the ozone layer as a problem for the North. They produce the CFCs. They get the skin cancers. It is up to them, not the South, to make the cuts. I think some of the big developing countries thought Vienna and Montreal were a bit of a con.'

The London conference

It was largely to overcome the suspicions and doubts of the developing countries that UNEP and the newly-greened British prime minister, Margaret Thatcher, mounted the 'Saving the Ozone Layer' conference in London in March 1989, an impressive gathering of ministers and delegates from more than 120 countries and well attended by academics and leaders of the CFC-producing industries. If it could generate the right mood of urgency and co-operation, it would ensure success for the Helsinki meeting of the Montreal Protocol parties in the following April. UNEP's executive director Mostafa Tolba, in a powerful statement that illuminated the road to salvation with visions of environmental hell-fire, warned that the alternative to a 'massive and sustained' global effort was catastrophe. 'We owe it to the newly born generation to do all in our power to hand over a living earth,' he declared. The British government's readiness to pay fares and expenses had encouraged a good attendance among Third World countries whose governments might otherwise have thought twice before spending on representation at an environmental conference. The choice of President Arap Moi of Kenya, host to UNEP's headquarters, as the opening speaker was intended as an assurance of the conference's bona fides, so far as the Third World was concerned.

The conference was preceded by a few days by two developments, one providing it with timely encouragement, the other with cause for added concern. The latter came from a Canadian research team at Alert Bay in Ellesmere Island, almost the last land before the North Pole. Their estimate was that ozone depletion over the Arctic during the winter would be of the order of 5 per cent. The Arctic does not duplicate the Antarctic in the severity of its climate or its isolation. It does not have an ozone hole. Nevertheless there were enough similarities, including the icy stratospheric clouds, to convince several teams of scientists that the loss of ozone could become severe, and eventually capable of harming the health of the large populations close to the high latitudes. The World Meteorological Organization announced in the April following the London conference that its measurements indicated that during the winters–springs since 1970 the total column ozone (in the troposphere as well as the stratosphere) had declined by more than 4 per cent between latitudes 30°N and 64°N (the 'vertical' distance between Cairo and Reykjavik, to give a rough guide). During the summer it had dropped by 1 per cent between the same latitudes. The decrease would have been more if tropospheric ozone (i.e. near the ground) had not been increasing by about 1 per cent a year during the last few decades.[9] There were at least some benefits from smog, it seemed.

The encouraging news was that the European Community's environment ministers had decided to ban all production and consumption of CFCs by the end of the century, a more radical step than that proposed by Britain and some other countries. This was followed the next day by a similar move by the United States. Among the major industrial countries at the London conference only the Soviet Union held back, the chairman of its committee for ozone protection, Professor Vladimir Zakharov, claiming that cyclones over northern Europe might have caused the thinning in the Arctic ozone layer. Soviet scientists were preparing their own analysis: 'Until we find a good alternative to CFCs, we will continue to use them,' he announced. China and India took a somewhat similar line, adding that there should be an international fund to finance research into alternatives to CFCs and pay for the free transfer of new technologies to developing countries.

Two months later at the Helsinki meeting or the parties to the protocol eight nations – including the Soviet Union, India and China – agreed without dissent to tighten up the protocol's timetable and phase out production and consumption of CFCs and halons by the end of the century. That decision was made binding at the June 1990 London review conference. China and India, which between them produce 80 per cent of the CFCs made in developing countries, gave a conditional assurance that they would sign the protocol by 1992, the year of the next review conference. Throughout the preceding 12 months they had made it clear in negotiations that they would co-operate only if they were satisfied by the arrangements for financial aid and technology transfers. Refrigerators, according to a Brussels official, have become 'the benchmark of progress' in both countries. It was only days before the conference that President Bush in response to a letter from Mrs Thatcher dropped US opposition to an international fund and thereby virtually assured it of success. The US and EC will sit on the fund's 14-strong executive committee, half of whose members will be drawn from developing countries. The fund will be managed through the World Bank and the UN Development Programme and will have $160 million at its disposal for the initial three-year period, with an extra $80 million when India and China sign, a development which will depend on their being satisfied that assurances on technology transfers are being honoured. China showed some sensitivity during the conference to US claims that it planned to produce 300 million CFC-coolant refrigerators – one for every household – by the year 2000. The actual figure, explained an official, was likely to be between 40 and 50 million, all of which would go to urban households as the 800 million people in the countryside were not covered by the programme.

The 1992 review conference will almost certainly see an acceleration of the phase-out of CFCs. But for the objections of the United States, Japan and the Soviet Union, the 1990 London conference might have swung behind the proposal supported by 13 countries, including three major CFC producers, that the deadline should be brought forward to 1997.

Montreal, the two London meetings and Helsinki had shown in the course of less than three years that the developed and developing world were not as far apart on at least one important environmental issue as some had feared. Fifty-nine countries had signed and ratified the protocol. Agreement was possible, although it still had to be seen whether it worked in developing countries desperately attempting to improve the qualities of life of their ever-increasing populations. CFCs are cheap to produce; the alternatives relatively expensive. Western industry has proved to be co-operative, with most of the producers of CFCs agreeing to work together on hydrofluoroalkanes (HFAs) with a low potential for destroying ozone.

The campaign to save the ozone layer is an encouraging story, but even the most optimistic account has to admit that the end of the tale remains uncertain. Even if CFCs and halons are phased out it will still take decades, perhaps a century, for their presence in the stratosphere to decline to what it was in the late 1960s and early 1970s.

Notes

1. *The Problem of the Stability of the Ozone Layer* (Atochem, Paris, 1989).
2. *Fluorocarbon/Ozone Update* (Du Pont, Geneva, 1989).
3. See *Action on Ozone* (UNEP, Nairobi, 1989).
4. See *A Major Statement* made by William H. Mansfield, Dep. Exec. Director of UNEP at the Sundance (Utah) symposium on climate change, August, 1989.
5. Fred Pearce, *Turning up the Heat* (Bodley Head, London, 1989), ch. 1.
6. Paper by Robert T. Watson of NASA at the June 1988 Toronto conference.
7. *New Scientist*, 27 May 1989 and 3 February 1990.
8. *Our Planet* (UNEP, Nairobi, March 1989).
9. Press release, WMO no. 434, 27 April 1989.

CHAPTER NINE

Climate change and agriculture: and a Soviet dissent

Russia's history can be read as a long battle with geography and climate. General Winter may have repelled invaders from Napoleon to Hitler, but he is a domestic tyrant from whom successive regimes have sought to escape by pushing south to more friendly climes and warm-water ports. Summer has never been a sufficient compensation for his excesses. It is a fickle, too-short season and for much of this century the Soviet Union's granaries have never been quite full enough for domestic and political comfort. This is despite the fact that its acreage of farmland is about the same as that of the United States, a nation that has become the world's greatest grain exporter. Tsarist Russia before the First World War held that position, earning a third of its foreign exchange from grain sales. Famine at home did not deter the Tsars from exporting grain; Stalin took a similarly hard-nosed approach, rationed consumption at home and sold abroad for the sake of political influence and foreign exchange. The Soviet Union managed to remain a net exporter until the beginning of the 1970s. From then on the contrast with the United States becomes stark. American production soared, largely to meet overseas demand, and the Soviet Union, despite some impressive gains in production, became for the first time a large importer. Lack of incentives, bureaucratic mismanagement and other well-publicized ills of the Soviet economy may bear much of the responsibility for the disparity between the superpowers, but Robert Paarlberg's *Food Trade and Foreign Policy* confirms that climate has been the chief factor:

More than half the arable land in the United States receives annually at least 700 mm of precipitation; only 1 per cent of the arable land in the Soviet Union receives as much moisture. In those regions where rainfall is adequate, the soil tends to be poor, or the weather too cold. Plentiful moisture in the extreme northern latitudes, for example, is rendered useless by severe winter frost and an abbreviated growing season. The warmer regions south and east of the Ukraine are unfortunately the driest. Light and irregular rainfall during the summer growing season here is often accompanied by high temperature, scorching wind, and devastating dust storms. Because the inadequate moisture levels are also highly changeable, year-to-year variation in Soviet farm output is typically three times greater than in the United States.[1]

Grain has always been important as an indicator of national well-being and prosperity, with bulging granaries more important to security in the long term than the missile silos which often share the same farmlands. The 1988 drought – the third of the decade – was arguably more of a blow to the American sense of security than the discovery of an alleged missile gap in the 1950s. Domestic consumption of grain overtook production for the first time. There are no recurrent natural events so mindlessly implacable and dispiriting as droughts and, regardless of the argument about the true cause of 1988s, the televised scenes of dying maize and despairing farmers became the hinge of a turn in the Western world's chief preoccupation, from the threat of nuclear annihilation to climate change. A new era had announced itself.

By contrast, the Soviet Union had, of course, always co-existed with a climate which could never be taken for granted. In the post-war period Moscow directed much effort towards overcoming its agricultural inferiority *vis-à-vis* the West, especially the United States. Stalin's successor Khruschev, in the mid-1950s, made overtaking US meat production (which entailed rapid increases in the production of feed-grains) the measure of success or failure of his attempt to improve the Soviet diet. He envisaged a Soviet return to the position of leading exporter of grain as something that would make 'Messrs Imperialists' sit up and take notice. In fact, Soviet production did rise until 1978, after which a run of six bad harvests and the inconsistencies of the Soviet economy produced stagnation. Given the poor publicity generated by the Soviet Union's agricultural failings, it comes as something of a surprise to look at the statistics and see that the Soviet Union is still the world's largest wheat producer, greater than the European Community (which also made agricultural self-sufficiency a goal in its early years) by about 10 million tons in 1989, outstripping the United States and Canada, the world's 'breadbasket', by almost the same amount. Where the Soviet Union falls down is in the production of what are called coarse grains – maize (corn) principally, but including barley, oats, rye, sorghum and other grains apart from rice. It was the maize crop which was the chief victim of the 1988 American drought. But even in a bad year like that one, US production was nearly eight times greater than the Soviet Union's. In a reasonably good year like 1989 it was between 12 and 13 times greater, fortunately, perhaps, since the Soviet Union was obliged to purchase 40 million tonnes of grain from abroad.

During the early 1970s, when the sort of thinking exemplified by *Limits to Growth* permeated official forecasts of imminent shortages of strategic commodities, the CIA toyed with the idea that 'The United States' near-monopoly position as a food exporter . . . could give [it] a measure of power it never had before.'[2] Food would become the weapon with which the United States would counter the Arab world's use of an oil embargo in its fight against Israel. It was an idea that persisted. Food had become power, a 'valuable instrument of foreign policy', as Alexander Haig, Secretary of State under President Reagan, noted in 1981. President Carter had applied it in the form of an embargo on grain sales to the Soviet Union after the Christmas 1979 invasion of Afghanistan; and it was to prove as ineffective as the Arab oil embargo.

The grain trade was – and is – of immense importance to the United States, and to its relation with the Soviet Union. It has entered into long-term bilateral grain agreements (LBGAs) with only two countries, the Soviet Union (three times) and China (once). During five years of the second LBGA (1983–8) with the Soviet Union, the United States sold its rival grain worth $7 billion and during two of those years – 1986 and 1987 – subsidized the sales to the tune of $450 million.[3] In the peak years of 1979–80 the United States' share of the world trade was 51 per cent of wheat and 73 per cent of coarse grains. It was a trade worth, in 1980, $27.1 billion. By contrast, the Soviet Union had become the world's biggest importer, taking almost a quarter of all internationally traded grain, a third of it from the United States.[4]

The Soviet Union is making strenuous efforts to reduce its dependence on imports, boosting its use of fertilizers to levels above those anywhere else, but its enlistment of poor, fragile land in the battle for self-sufficiency in the early 1970s has meant a loss of land through abandonment in the 1980s, perhaps as much as 13 per cent of the peak grain area.[5] Food may not have become a weapon, but the failure of Soviet agriculture to meet its peoples' demands was a wound that contributed to the collapse of Soviet power in the second half of the 1980s.

Climate change could put both superpowers in the same boat. There are those who see the drying out of the continental interiors as a more significant consequence of climate change than sea-level rises. Environmental politics on the global scale will be dominated increasingly by the need to feed people, so any climate shift adversely affecting the American grain-belt will be a serious matter. Other areas may, of course, be able to produce more. Agricultural adaptation is possible, too. The Israelis have grown fruit and nuts successfully in temperatures of 46°C and new strains of drought- and heat-resistant crops will no doubt be developed. How quickly, though, and whether there will be sufficient incentive to produce in bulk for countries which are unable to pay are different matters. Countries like Ethiopia have been fed out of surpluses at little cost, but even there thousands have died before the food reached them. In much of the West where productivity has been high and population increases relatively small – in some countries zero – the problem confronting agriculture has been how to control expensive surpluses. Butter mountains loom above lakes of olive oil.

Milk quotas have been introduced to limit the output of over-productive herds. Set-aside schemes have taken land out of production in the United States as well as in the European Community. The United States, for instance, idled 15 per cent of its total crop acreage in 1987 and 1988 in line with the Administration's policy of reducing stocks, raising prices and conserving land at risk from erosion. One of the consequences of the 1988 drought was a sharp reduction in the set-aside programme. Lester Brown, the head of the Worldwatch Institute in Washington, DC, calls the food mountains of the 1980s a 'temporary aberration' which will soon run out.[6] The 1980s, he notes, were the first decade in history in which there was no increase in the area of land farmed and – also for the first time – a fall in the amount of fresh water used per capita. Shortages and rising prices could mean the world was heading for 'the grain shock of 1992', a crisis for which it was unprepared.

Towards a brave new modified climate

Mikhail Budyko, the most revered figure of Soviet climatology, has never been noted as a cold war warrior, but he has always been aware of the limitations imposed on the Soviet Union by its climate. He is not a sceptic about the reality of global warming – he can claim to have been the first man to notice it – but he does not believe it is necessarily harmful. He accepts the evidence that there has been a drying-out in the continental interiors affecting both the United States and the Soviet Union and then qualifies his acceptance by suggesting that it will be remedied in the next century by greater precipitation. He uses a study of past periods of warming stretching back over several millions of years to confirm his view that more carbon dioxide to nourish plants, more warmth, more rainfall, more arable land as the frontiers of ice are pushed back will mean bigger harvests. His thinking, with its faith in science and brave new worlds, is very old-style 'Soviet', at its core a belief that climate can be modified beneficially on the grand scale. He recommends more research, with 'the aim of creating a global system of biospheric regulation in the interest of human society'.[7] In *Climate and Life* he recalls the work of his collaborator O.A. Drozdov on 'Changing the Climate in Connection with the Plan for Transforming Nature in the Drought Regions of the USSR'; and studies by others on the feasibility of melting the polar ice with nuclear energy and by damming the Behring Straits and pumping water from the Behring Sea into the Arctic. It is by no means evident, he admits, that the results would be beneficial; more study is required. He is more convinced by a scheme prepared by another collaborator, M.I. Iudin, for changing the intensity of low-level cyclones by employing a system of vertical air jets in the area where they form. Laying down broad belts of asphalt might, by developing thermal convention, lead to increased precipitation. He writes in the last paragraph, 'It should be borne in mind that for the creation of a climatic regime managed by man, further progress of science and engineering is necessary which would permit a considerable increase in the present production of energy. It is beyond doubt that with peaceful development of human society, such progress will be realised in the near future.'[8]

'Budyko thinks climate change is the best thing since sliced bread,' said a US scientist at the World Meteorological Organization in Geneva in 1989 in a conversation with the author. 'He is greatly respected, but I don't think the Soviet-led "impacts" group [of the IPCC] will reflect his views.' Others echoed his opinion, with differing degrees of confidence. The belief in a scientifically-controlled world runs deep in Soviet society. If you can't escape your climate, change it. If you don't think your rivers are running in the right place, divert them. The ecological disaster created by the diversion of water from the Aral Sea for the sake of a better cotton crop has given that sort of thinking a severe knock. The Aral has lost 40 per cent of its surface as well as its fish and the fishing fleets that netted them. Salt-laden winds howl round its desiccated and still shrinking perimeter. But in a world where nature left to its own devices is as grim as it is in much of the Soviet Union, the belief that you can bend it on the heroic scale is unlikely to go away. Climate change through the man-made modification of greenhouse gases might be disastrous, but it might equally be deliverance. There might even be, one day, warm water ports in the Arctic. Soviet hesitations were evident when Budyko told a climate change conference in Hamburg in November 1988 that it might be better to increase carbon dioxide emissions to encourage warming. The cost of restricting emissions in a country struggling with inefficient smokestack industries may not have been in Budyko's mind, but it is an obvious factor in the Soviet attitude, as it is elsewhere.

Obviously, there is always a temptation to think nationally rather than globally. If India's government is persuaded that warming will produce better and more reliable monsoons, then it might decide that the interests of its burgeoning population would be better served by global warming than by attempts to hold it in check. Even Canada, a leader in environmental matters, must have interest in a situation which could extend its growing season and allow it to push its arable lands further north. But the trouble with climate modification of the sort imposed by the greenhouse effect is that it does not move naturally to a new, warmer status quo which will remain undisturbed for the next thousand years or so. It continues to change, and probably very rapidly, so any gains might well prove to be short-lived. In large nations and groupings of nations, such as Western Europe, advantage in one area might be accompanied by misfortune in another, as, for example, if Northern Europe experienced a more favourable climate while the Mediterranean dried out.

The decade of the 1980s had its upturns as well as downturns in food production, but the overall pattern was disquieting. The three drought years sapped North American output. From the middle of the decade onwards the two largest Third World countries, China and India, failed to increase their grain production. It was clear from the records of yields per hectare that the era of 'miracle' grains was over. India, which prided itself on the self-sufficiency achieved by its green revolution, was obliged to buy wheat in 1988–9; China's imports of wheat reached record levels in the last years of the decade, a reflection, no doubt, not just of two years of drought but of a situation where peasants received certificates instead of cash for their crops. Poland, a developed country, joined the ranks of supplicants for food aid, a category usually limited, under the Food Aid Convention, to developing nations. Despite a 6 per cent increase in the global cereal harvest in 1989, the year was the third in succession in which production failed to meet demand. Food stocks fell to their lowest level since the food crisis of the early 1970s. How accurate such figures are is always open to question, though. Local farmers and regional government often hold large emergency stocks which are not counted, but it is nevertheless accepted that they represent a trend.

The figures for food sent by the 22 Food Aid Convention countries reveal once again the importance of the US grain belt. It provides more than half the aid, 75 per cent of which is wheat. So a cut of 30 per cent in the US contribution of wheat is immediately reflected in a similar cut in the total amount available. The effect of that in 1988–9 was not starvation in developing countries but higher food bills. Whereas in 1987–8 some 14 per cent of their wheat imports came as aid, in 1988–9 the percentage was down to 11 per cent. 1989–90 looked like being much worse, with the percentage falling to 8 per cent. As the International Wheat Council noted in its October 1989 market report,[9] the implications for countries struggling with foreign debts, higher commercial grain prices and tougher credit terms were 'especially serious'. On the other hand – and there is invariably an 'other hand' in these environmental and quasi-environmental matters – aid tends to depress local markets and discourage farmers from producing more. Reduce aid and they will grow more, goes the argument. That may apply to some African countries, but it is hard to imagine the two biggest aid recipients, Egypt and Bangladesh, doing a great deal more to feed themselves.

No more 'miracles'

Abundance often lies less than half a day's jet flight from starvation. Distribution problems have probably caused more deaths from starvation in recent years than an actual shortage of food. The West could certainly produce more than it does, but equally certainly not enough to deal with the huge population increases projected for the developing world in the coming century. At the June 1988 Toronto conference three members of the Indian Agricultural Research Centre, in New Delhi,[10] provided some figures which demonstrated the size of the problem. In 1986, 1942 million tonnes of food grains were produced to feed a world population of nearly 5 billion. To maintain the same level of per capita consumption in 2025, with an expected population of 8.2 billion, grain production would have to increase to 3050 million tonnes. To take a few regional examples, Africa would have to increase grain production by almost fivefold and Latin America by almost threefold. Apart from Africa, there is not a great deal of potential arable land left, so, using an across-the-board figure, it would be necessary to devote approximately 66 per cent of all farmed lands to cereals. Even so, the bulk of the increase would have to come from improved yields, as it has in the past three decades.

At that point, two factors slide into the picture. The first is the energy, most of it from fossil fuels, required by modern agriculture and the system which delivers its produce to the customers; the other, the peaking of the 'miracle grain' revolution. New and more productive cereal cultivars have contributed their share to improved yields, but the biggest share in developed countries like the United States has come from increased use of fertilizers – 55 per cent between 1965 and 1976, according to the New Delhi paper quoting an FAO estimate. Modern agriculture requires not only more fertilizer but more mechanization, too:

> This in effect led to a situation where the commercial energy input became about equal to the equivalent energy output of edible grains. If the energy input to the whole agricultural system was estimated, from land preparation to canned provisions in supermarkets, then the commercial energy expended was greater than the solar energy harvested by the crops. Thus it would be virtually true to say that ultimately fossil fuels serve as food in developed countries.

The *State of the World 1989* put the same case in a different form:

> In per capita terms, world fertilizer use quintupled between 1950 and 1984, going from 5 kilograms to 26 and offsetting a one-third decline in grain area per person. As land becomes scarce, farmers rely more on additional fertilizer to expand output, in effect substituting energy in the form of fertilizer for land in the production process.

Farmers in developing countries use fertilizers on their lands, too, of course, but more of the routine work is done manually. The New Delhi paper estimated that 0.038 of a ton of oil was burnt to produce a ton of grain in India; in the United States the figure was 0.110. More fertilizer might mean more grain (and more carbon dioxide and nitrous oxide to add to the greenhouse effect), but there obviously comes a point where more fertilizer does not mean a bigger yield or perhaps is not justified on cost grounds, particularly when grain prices are low, as they were for part of the 1980s. The miracle grains introduced in the 1960s have played their part in the agricultural revolution but there is no sign that new strains are on their way. According to the New Delhi paper's authors, the peak yields in both wheat and rice were obtained during experimental trials in the 1960s and have not been repeated since. So any increase in output will have to come from better farming methods, particularly in countries like India, whose rice farmers lag well behind Japan's in productivity.

The growth of the deserts

Human ingenuity has often confounded the pessimists. Malthus was wrong about the effects of unchecked population growth, as the optimists constantly remind those who look too bleakly into the future. The law of diminishing returns has not applied to food production in the countries which have undergone the industrial revolution, although there have been periodic crises, such as the Irish famine of 1845–6, to remind us that Malthus was not completely wrong. About one-third of the Irish population emigrated to wrest a living from new lands in North America and Australia. Mass emigration of that sort is not an option these days. Leaving out the debatable new lands opened up by clearing tropical forests, the amount of land available for agriculture, even of the most limited, nomadic sort, seems to be diminishing. According to UNEP's figures, desertification is taking as much land out of agricultural use as exploiters of one kind or another are clearing in the rainforests:

> UNEP and others have estimated that it would cost approximately $4,500 million a year over the next 20 years to slow and stop desertification, and to begin to reclaim the land that we have already lost. That money and other needed support just hasn't been forthcoming, and because of that the deserts continue to advance. Pilot projects don't get catalyzed; know-how isn't spread among the very poorest countries; outside development assistance continues to be poorly co-ordinated – sometimes even pulling in opposite directions.[11]

Desertification comes in several guises: as encroaching sand, degrading croplands and grazing, waterlogging and salinization of irrigated land, destruction of trees and shrubs, erosion and deterioration in water supplies. It is above all largely caused by people, with nature in the accomplice role. Under threat are the world's drylands, which are said to constitute about 35 per cent of the world's land area and support some 850 million people. Tolba accuses the international community of neglecting problems which will grow worse with time: 'The real obstacle is the lack of political will.' There are others, though, who say that UNEP's Plan of Action is largely irrelevant to the attempts to stabilize the drylands. Most aid is being given in the form of bilateral projects. And by what method of assessment (ask the critics) did it arrive at its figure of $90 billion for a twenty-year programme? The truth is that such huge figures have a hypnotizing effect on the victims and arouse a defensive incredulity among those supposed to provide the money. The victims are unlikely to believe that they can do much from their own resources if action is dependent on billions of dollars, and the developed nations can hardly man every environmental barricade from Ethiopia to the Amazonian rainforest. There may be, too, a *sotto voce* challenge to the received wisdom that it is people who cause desertification. Deserts are the result of a lack of water and if there is a prolonged, fluctuating drought (like the one that began on the northern fringe of sub-Saharan Africa at the end of the 1960s) vast sums of money are not going to stop desertification.

Where desertification poses its greatest threat to the lives and livelihoods of people is in the Sahel and the Horn of Africa. It is not a region where costly Plans of Action are going to work. 'There isn't much point in trying to grow a belt of trees across the desert,' Camilla Toulmin, the head of IIED's Drylands Programme, told the author. 'I would make two points: the first is that a lot of very small-scale activity does achieve results. And the second is the need to get people involved locally. I wouldn't say the desert is being contained, but there are diagnoses and solutions.'

The solutions in the end come down to not over-grazing, managing the water supplies and choosing appropriate crops. But sustainability in such arid circumstances depends on sustainable populations which do not exceed the resources. And in Africa that must mean people moving from the rural areas into the overcrowded towns, which at least offer subsistence. They also offer political instability. As is frequently pointed out, every government in the Sahel fell after the droughts of the early 1970s. Poor, vulnerable government makes every problem worse, particularly in a continent where so many factors contributed to increasing poverty and very few to wealth. As a World Bank report[12] noted grimly, 'Never in human history has population grown so fast.' The half-billion people in sub-Saharan Africa will become one billion by 2010, a growth rate which, now or later, has disaster as its terminus. Political will in such situations is a low explosive, blasting along the lines of least resistance. Alas, there is no profitable course of least resistance in Africa.

If one takes everything into account – the shifts in climate which might well remove the United States from its place as the leading 'food power' and affect the Soviet Union as well, the limited prospects for increased yields, desertification and the steady increase in the global population – it is clear that in the next century large sections of the world's population will have a very narrow margin between them and starvation. In an increasing number of cases, no margin at all, as has been the case in the past and is so today in parts of Africa. There are cushions like the World Food Programme and the stocks held by the major grain producers, but emergencies and poor harvests could run them down very quickly. But there are no easy solutions; it is hard, for example, to imagine the EC countries deciding overnight to forget their searing battles over agricultural policy and return to the old days of costly subsidies and surpluses in order to build up buffer stocks to higher levels. It is equally hard to imagine countries capable of producing surpluses stonily refusing to do so in the face of an evident need. At the moment, though, understanding of the regional impacts of climate change really is not good enough to persuade countries to rush into costly agreements on strengthening food stocks. Countries can think about widespread famine and the international organization needed to deal with it, but it is not easy to make detailed plans years in advance of the crisis. The 1992 UN conference on the environment could be the right place for an analytical look at what the world needs to do to ensure that it can feed its vulnerable populations in the next country.

Notes

1. Robert L. Paarlberg, *Food Trade and Foreign Policy: India, the Soviet Union and the United States* (Cornell University Press, Ithaca, New York, 1985), p. 69.
2. Report by Central Intelligence Agency on *Potential Implications of Trends in World Population, Food Production and Climate*, quoted in *Food Trade and Foreign Policy*, op. cit., p. 13.
3. General Accounting Office, Washington, DC, April 1989. GAO/NSIAD-89-91 (*International Trade: Long-Term Bilateral Agreements and Grain Counter-Trade*).

4. Nick Butler, *The International Grain Trade: Problems and Prospects* (Croom Helm, London, for the Royal Institute of International Affairs, 1986).
5. *State of the World 1989* (World Watch Institute, W.W. Norton, London), ch. 3.
6. See *New Scientist*, 31 March 1990.
7. *Ecologie globale* (Éditions du Progrès, Moscow, 1980).
8. *Climate and Life* (Academic Press, New York, 1974).
9. International Wheat Council Market Report, 26 October 1989 and *State of the World 1989*, op. cit.
10. S.K. Sinha, N.H. Rao and M.S. Swaminathan, *Food Security in the Changing Global Climate*.
11. Quoted in UNEP's *Regional Bulletin for Europe*, June 1987.
12. *Sub-Saharan Africa, from Crisis to Sustainable Growth* (The World Bank, Washington, DC, 1989).

CHAPTER TEN

Eco-politics: The greening of the greys

The Cité des Sciences et de l'Industrie near the Porte de la Villette in Paris is one of those technological markers the French like to put down to show that architecture in their capital is not all ormulu and *Belle Epoque*. There is also concrete, steel tubing and glass to think about, materials judged more suited to framing the aspirations of modern man. As the venue for the fifth international conference of the European Greens, the Cité, with its predications on a science-led industrialized society, seemed about as sympathetic a choice as the Chicago stockyards for a vegetarian convention. The Greens had planned to hold their 1989 rally in the Sorbonne, but at the last minute they were told they would have to go elsewhere and the only venue available at short notice was the Cité. The hand of President François Mitterrand was detected. Euro-elections were pending and, it was rumoured, he was doing his best to make life miserable for an organization which challenged his claim to the green constituency. 'Give the meek a good deal – join the Greens,' announced a poster on a publicity stand in the foyer of the conference hall. The 1000 or so members of the meek class who presented themselves at the conference, were, suitably enough, unruffled by having been shunted from the Sorbonne to the techno-chic sidings at Porte de la Villette. Inheriting the earth, or what's left of it by the time the non-meek have realized the folly of their ways, is a process which takes much fortitude and patience.

It is, in fact, the absence of pushiness which makes the

Greens a surprising lot, essentially more religious (in the very broadest sense) than political; more activist Quaker than Marxist. The British Greens, for example, allow time for 'attunement' before their meetings start, reminiscent in its way of the reflective silence that can prevail at meetings of the Society of Friends. The representatives from seventeen national parties at the Paris conference were quite plainly non-attenders in the established church of politics. No priests, merely a shifting scene of spokespeople and co-chairs, and no franchise held by anyone on the pulpit. The Greens are officially leaderless. When the light descends on the individual and the voice within speaks it has an equal right to be heard. It is the individual and the change of heart which prompts him to participate which matters. Only the Germans, with their endless internal quarrels (and who are perhaps more faction-ridden Buddhist than Quaker in their attitudes), challenge the rules of green decorum. They come from the alternative left originally, a sector which at its extremes is tinged with violence. But one can go on almost endlessly tinker-tailoring the sources of greenness: Nordic nature-loving fascist, Quaker, Zen Buddhist, Taoist, Social Creditist, Marxist, socialist, pacifist, quietist, 1930s Green Shirt. John Hargrave, the Quaker founder of the Kibbo Kift Kin, came nearest to bringing it all together when he began his book on the Kin with a quotation from the Taoist Lao-Tzu: 'production without possession, action without self-assertion, development without domination.'[1]

The fact that no one is given command in the green movements does not mean that no one stands out among the ruck. There are spokespeople (or speakers) and individuals whose characters and strong views shape the movement: Sara Parkin, who lives in Lyons and is co-secretary of the co-ordinating body which links the European parties; the Swiss Daniel Brélaz who, in 1979, became the first Green to be elected to a national parliament; Petra Kelly, the uncompromising driving force in the Fundi (fundamentalist) faction of Die Grünen, whose pragmatic factions, the Realos (realistic reformers) and Superrealos, seek to establish working relationships with the mainstream Left; and a number, not a very large number, of others.

The Greens are, of course, tributary left, at times in spate, at other times the merest trickle, all according to the refreshments offered by fickle public opinion. Their course is so far short: the first national Green Party founded in New Zealand as the Value Party in 1972, the next in the United Kingdom in the following year and then a long gap until Ecolo in Belgium in 1980 heralded a rush either to set up or amalgamate existing groups into parties in the early and mid-1980s. Seen and heard *en masse* those at the Paris conference were recognizable as the middle class's awkward squad: idealistic non-makers of money, profoundly suspicious of governments and the works of capitalism, individualistic and a touch anarchistic, decent in a stubborn, purposeful way and kindly. They are the sort of people (and very often the same people) that in Britain you meet in CND and in scores of cluttered offices dedicated to under-regarded and under-rewarded causes.

A look later at a partial list of the Green Party's candidates for the 1989 British Euro-elections confirmed the general impression: it was full of teachers, lecturers, organic small-holders, social workers and adult students, with here and there the stray doctor, engineer and journalist. A husband and wife duo of lawyers-turned-smallholders may have founded the British Green Party, but there were no lawyers on the list, or surgeons, or stockbrokers; or, for that matter, plumbers or shop stewards. They are not a noticeably intellectual lot, the Greens, and anyone who goes to one of their conferences expecting deeply researched papers on the carbon tax issue or the structure of the future Green confederation of Europe will come away disappointed. It is feelings, the heart not the mind, which are at the core of the Green polity. As so often with essentially good people who act from feelings, they are broadly right. What is wrong with the Greens is that they do not acknowledge the profound economic difficulties, and all their attendant national and international jealousies and suspicions, which obstruct the path to an environmentally safer world. They suffer, in short, from that most unforgiveable of political sins, unworldliness.

The Euro-elections were something of a test for the Greens. They were fighting for seats in a body – the Strasbourg

parliament – which is the assembly of the European
Community, of which they strongly disapprove. 1992 and the
European Single Market would be 'disastrous, catastrophic',
an 'ideological bluff' devised by deceitful politicians, declared
a German MEP at the Paris conference. 'The new material
culture will mean McDonalds from Sicily to Scotland.' On the
same theme, the UK party made plain in a pre-European
election statement why it did not like the Community:

> The European Community consists of nation-states
> united by their desire to create greater economic wealth
> for themselves. This is incompatible with the constraints
> of a finite world and we are therefore opposed to the EC
> as presently constituted.

The reference to a finite world carried echoes from
Blueprint for Survival, the founding document of the British
party in the early 1970s. The party's formal manifesto (with a
foreword by Petra Kelly) was no less dismissive, predicting
failure for the 'European superstore'. It is not just the EC's
free market consumerism which the Greens detest; they are
opposed, too, to its political structure based on capitalist
nation-states. The Greens believe (said the British manifesto)
in a decentralized 'Europe of the Regions' which would
develop into an all-embracing confederation 'from Lappland
to Armenia'. How that would work out in practice is unclear.
In Britain, there are Scottish Greens and Welsh Greens, but
no English Greens. All managed somehow to merge their
identities to produce a Euro-election manifesto in the name of
the Green Party, which is the generic name under which the
party appears on the voting slips. Diminutive Belgium is more
precise: its Greens are divided between Flemish-speaking
Agalev and French-speaking Ecolo.

Greens are against multinationals, naturally, and even
against the ending of monetary barriers within the EC because
'wealth will migrate even more quickly to those areas which
are already wealthy, while the "peripheral areas" will
continue to be fed grudgingly with money for inappropriate
projects like wall-to-wall holiday apartments and nuclear
waste dumps.' (The leisure industry was denounced during the

Paris conference as 'prostitution'.) The Greens want regional self-reliance, in agriculture as well as in other things; a land tax applied so that 'in general terms, the nearer the land is to its natural state, the lower the land tax would be'; energy efficiency; population reduction through encouragement and education (with 15–20 million the target for Britain); and a sharing of 'the abundance which nature can provide for us all if the greedy do not take more than their fair share'. It was easy to make fun of, and it was no wonder that Conservative Central Office had a field day, distributing a compilation of green policy statements, including the unrealistic ones on replacing the EC with a 'loose federation of ecological countries' and disengaging Britain from the international money market. The Conservatives handed out the Greens' fairly innocuous version of the 1989 Queen's Speech gratis, too, in the belief that it was time the public had their eyes opened to what the party was really about.

The conflict between abhorrence of the institution and the political expediency which obliges Greens to seek seats in it leads to a rather antiphonal quality in much of what the Greens have to say on Europe and the devolution of power to the regions. Their belief in 'maximum self-reliance' with regional parliaments retaining 'most legislative powers' did not prevent the British manifesto from declaring that a Green government would use the European courts to speed up UK compliance with EC environmental directives. Equally, there would be a 'coherent, integrated, efficient European transport system based primarily on waterways and railways', but no Channel Tunnel. That presumably would remain the English 'region's' position on environmental grounds even though the French 'region' on the other side of the Channel might favour the tunnel in the interests of an integrated railway system.

Even presented in outline like that, it is reasonably clear that under green regimes Europeans would, by any conventional standards, become a great deal poorer than they are at the moment. The Greens admitted that implicitly in the 1984 Joint Declaration of the European Green Parties in which they rejected an 'economy based primarily on productivity [and] . . . on the creation of artificial needs'. A better way of life need not depend on a higher standard of living and they

were out to stop those whose pursuit of continued economic
and industrial growth undermined the basis of life itself. 'We
wish to break totally from the liberal capitalism of the West
and the state capitalism of the East, and want a third path
which is compatible with an ecological society.' Deep within
such essentially non-economic thinking is an image of a
bucolic arcadia where demands and ambitions are simple and
limited, the Amish of Pennsylvania on an international scale,
but with railways and EC directives on pollution. The
unfortunate reality is that Europe is heavily urbanized and
densely populated and there is no prospect, short of disaster,
of changing that in the foreseeable future. The Greens may be
the wise visionaries of Europe, but in a 'serious' election
dealing with economic issues their manifestoes would be
quickly torn to shreds. No one votes for poverty – at least, not
yet. So why did the British Greens (to take the national party
which won the highest percentage of the vote) do so well in
the 1989 Euro-elections?

First, because Euro-elections do not have the blood-letting
ferocity of a general election. What happens in Strasbourg is
not going to have much effect on the mortgage rate or
inflation. The election coincided with a precipitous slide in the
popularity of a Conservative government struggling with a
severe balance of payments deficit. It was a great opportunity
for registering resentment. Taking a gamble, the British
Greens contested all 78 mainland seats and took nearly 15 per
cent of the vote (but no seats because of the first-past-the-post
rule of British elections; i.e. there was not a single
constituency in which the Greens had a majority). In the 1984
Euro-election, when they contested only seventeen seats, they
took 0.2 per cent. So it was a remarkable jump into the
headlines and popular acceptance, even if it was of a rather
limited nature. It was true that the turnout was a meagre 37
per cent of the electorate, by far the lowest in Europe, where
the average was 59 per cent, but it was still two million
people. The heaviest green voting was in the Tory south,
lending weight to the belief that many voters were dis-
enchanted Tories casting a vote against an unpopular
government. It was a view supported later by the fading away
of Green support to between 3 and 5 per cent in polls held at

the end of the year. The voters were by then more hostile to Mrs Thatcher and her government, but they were switching their support to Labour, not to the third parties. The hard core of support seemed to have risen slightly during the year, but voting Green in the Euro-elections had not for most of the voters signified a new allegiance. Relatively few had bothered to examine Green Party policy in detail, and most were unaware that the party was hostile to the EC and proposed reducing consumption in the interests of conservation.[2]

The protest vote theory, though valid, was not sufficient to explain why 2 million adults took the trouble to vote for organic smallholders and kindred green souls who had hitherto made little impression on the political scene. The vote was overwhelmingly urban, as it must be in a densely populated country like Britain, and almost certainly middle-class. In similar circumstances in previous elections the dissatisfied Tories would have given the Liberals and the Social Democrats the benefit of their protest. So their switch to the Greens indicated something more positive in their choice than mere protest. Put at its simplest, they were voting for the countryside. The fact that the Greens' 'Europe of the Regions' is little more than a nostalgic fantasy and their economic policies might rapidly place even organic small-holders in the same income bracket as Third World peasants was not important. The Green voters were supporting clean air, unpolluted rivers, a peaceful countryside unblemished by motorways and new estates, in short for 'quality of life'. They voted for a restoration of a betrayed heritage of wild flowers and healthy trees. It might be as much a shimmering dream as 'Europe of the Regions' but that does not diminish its power. A Gallup poll conducted for the *Daily Telegraph* two months after[3] the Euro-elections found that an 'overwhelming' majority of the electorate, including three-quarters of those who voted Tory in the 1987 general election, believed the government was not doing enough to protect the countryside. More people than ever wanted to live there rather than in towns and 79 per cent took the view that the countryside was in danger.

Reviewing the situation at the end of the year, Lindsay Cooke, one of the Greens' 1989 spokespeople, thought despondently that the movement's share of the poll had shrunk because of a lack of publicity. There were no MPs in Strasbourg or in Westminster for the media to focus on. 'The media have decided that there has been a return to two-party politics and everyone else has been marginalized,' she told the author. 'Our views are not sought and when we do make our views known, they are not interested in reporting them.'

The number of local Green parties had risen to nearly 270 by the end of the year and the membership had almost doubled from 9000 at the beginning of the year to 17,000 by the end. Support for the Greens may have shrunk but it was still at the same level as the Liberal Democrats', the third party in Parliament, so perhaps Ms Cooke was unduly downcast. But despite those substantial gains, bold statements about fighting every seat in the next general election were no longer heard. The party was short of funds; and it was still being run from a tiny office with just four paid staff. A third of the membership claimed to be unemployed and their reduced subscriptions of £6 did not cover even mailing costs.[4] The press officer, Barbara Bloomfield, complained at a meeting of the Greens' council in January 1990 that if she wanted to step out of the office she had to get a volunteer to take her place, possibly an overseas student who knew next to nothing about the press or the party. Friends of the Earth, by comparison, had ninety paid staff. 'We cannot continue to run a serious party like this.' Chain-stores and industries may take full-page advertisements in newspapers to proclaim their greenness, but no millionaire had come forward to offer a share of his eco-sound profits. If a potential philanthropist did exist, he might have found it a deterrent that there was no established leadership to whom he could talk. The Greens were still arguing over whether the time had come to loosen principle a little and elect a leader, a public face, perhaps Sara Parkin, the International Liaison Secretary of the UK Greens as well as the European Greens' co-ordination co-secretary. 'We believe in leadership *with*, not leadership *over* the party,' said Lindsay Cooke. 'As a problem, it's a media invention, and, as it happened, the media chose a leader for us – Sara Parkin. I am not decrying Sara in any way when I say it might have been more sensible for *us* to choose the leader.'

Two million votes and no seats. The British electoral system presents practically insuperable obstacles for small parties to overcome. Under proportional representation – the rule of all other EC member states – the British Greens would have been the largest national group in the Strasbourg Green firmament, where the number of those who regarded themselves as in some way Green had risen from 11 to 43. That position was occupied by the French (whose vote had also risen steeply) with nine seats, followed by the West Germans with eight and the Italians with seven. There were in the parliament at the end of 1989 twenty-nine European Greens and fourteen members (including two defectors from the Greens) of their old partnership, the Rainbow Group, a 'technical'[5] gathering of green-tinged communists, nationalists and anti-EC Danes. Even after this hiving-off the European Greens were not exactly monolithic, since, apart from the ideological differences within Die Grünen, they had taken under their wing an Italian who wanted the legalization of drugs and several small, otherwise homeless parties.

The most dynamic Green party in Europe is undoubtedly Die Grünen, which took 8.3 per cent of the vote in the 1987 West German federal elections and won 42 seats, only four fewer than the ruling Christian Democrats' coalition partners, the Free Democrats. Here, if anywhere, one would imagine, was the making of the Red–Green coalition with the mainstream left-centre that the Realos seek. Such coalitions exist at the municipal level in West Berlin and Frankfurt. Informal talks were held with the Social Democrats in June 1989 with the 1990 elections in mind, a recognition in itself that the Green pragmatists were in the ascendant and the Marxist regionalists on the wane. And then the Greens' past caught up with them – or, more accurately, was put on display by the government parties – and the Social Democrats retreated rapidly in case the electorate took fright at the thought of a pact with such a dubious, anarchic bunch. Die Grünen has been described by Petra Kelly as an 'anti-party party', and getting into bed with them might be a guarantee of ending up on the floor. They are the 'watermelon party', green on the outside and red on the inside, part of the 'alternative' subculture, the libertarian siblings of a rich society, not unlike at their extreme end the American drop-out society of the 1960s and 1970s. The communist Greens insist that Die Grünen's posture in the Bundestag must be one of 'irreconcilable opposition to and critique of existing conditions'. And there was more than a smack of sympathy for terrorism in a call by a leading Fundi for a 'broad show of unity' with a group which shot dead two policemen during a demonstration at Frankfurt airport.[6] Given the extremes within the party it is not surprising that there is a continuing feud between those willing to make political deals and those so utterly opposed to the system that they will make no concessions. Die Grünen is generally regarded as the most turbulent and self-destructive of the Green parties, but its internal quarrels are, says Sara Parkin in her guide to the European Greens,[7] 'only a more flagrant example' of what goes on in all the parties.

The Greens may have acted as catalysts in the green revolution, and their success in the Euro-election undoubtedly awoke the mainstream parties to the extent of the stirrings in the electoral loam, but they no more held a patent on green ideas than feminists did on equal rights for women. Their ideas could be expropriated as freely and easily as black-berries from a hedge in summer. For Labour, accommodation to the changing mood meant adopting the essentially green idea of assessing the environmental costs of growth and working them into the balance sheets (which the Tories are also adopting), but it did not mean following the Greens in calling for a halt to growth. To do that in the way the Greens proposed would require a 'virtually totalitarian state', Bryan Gould (at the time Shadow Trade Secretary) told the shadow cabinet in July 1989.[8] He admitted, though, that Labour was seen as too producer-oriented: 'We should be able to give a value to unpolluted air, clean water, a safe environment, as the end-products of economic activity.'

A great many green ideas were produced by Labour that summer; a 'green bill' to show up the deficiencies in Tory legislation; making the quality of life the central theme in the 1990 local elections; the mobilization of public opinion through the development of a network of environmental, countryside and energy organizations. The Trades Union

Congress weighed in with a Green Charter, which included the right to hold 'green strikes' over issues like the importation of toxic wastes. But by the end of the year the 'green bill' had been quietly dropped and the environmental 'network' had vanished into a general belief in the need for good, informal relations with Friends of the Earth and other environmental organizations. It was no longer necessary to upstage the Green Party; and it was becoming harder, too, to find a straightforward, easily defined green issue on which to challenge the Tories. Mrs Thatcher had established her environmentalist credentials in several speeches and had replaced the true-blue but ungreen Nicholas Ridley with the practically viridescent Chris Patten as Environment Secretary.

Labour was obliged to match Patten with Bryan Gould: 'I would agree that Thatcher and Patten understand global warming,' said Gould's environmental researcher and adviser, Nigel Stanley. 'Thatcher's speech to the UN [8 November 1989] was very effective. But I wouldn't agree that there is very little difference between the parties on green issues. The Tories may talk green but they act dirty.'

'Dirty' in this context means trying to apply free market ideas to the prevention of pollution, to continuing to dump sewage in the sea, to not doing enough to curb road traffic and improve public transport. Under the heading 'A better quality of life', Labour's policy review for the 1990s[10] declared that the future of the planet depended on keeping it at the top of the agenda:

> People want tougher environmental regulation – nationally and internationally. Industry, agriculture, local authorities and commercial developers must all have a clear framework within which to plan and invest.

It is hard to define precisely what constitutes quality of life. You know it when you experience it, but it is not the same thing as standards of living or GNP; it is not quantifiable and not necessarily 'environmental'. For some it is cows in flowering meadows, for others, cars and cheap central heating; and for the great majority, both. The quality of life issues around which so much of the green argument revolves are not trivial, but they are secondary. It is always possible to be kinder to the earth, but it is not saving clumps of trees from the bulldozers which will matter in the next century. The primary issues are industrial rather than country matters and concerned with survival rather than the quality of life. But if you strip away the embroidery from the Labour statement, there is agreement that the issue is survival. That at least seemed to be bipartisan. 'It is life itself we must battle to preserve,' said Mrs Thatcher in her speech to the United Nations in New York. 'The evidence is there. The damage is being done.' The head of government of any industrialized state, of whatever persuasion, could have echoed that fervently in 1989.

Notes

1. *The Confessions of the Kibbo Kift. A Declaration and General Exposition of the Work of the Kindred* (London, 1927). See Anna Bramwell, *Ecology in the 20th Century, a History*, ch. 6, for an excellent account of Hargrave and his contemporaries.
2. *Independent/NOP* poll, 7 July 1989.
3. 14 August 1989.
4. *Independent*, 20 January 1990.
5. Small parties are obliged to form a group of at least ten MEPs from at least three countries if they wish to qualify in Strasbourg for seats on commissions, research and office staff and the right to appoint a speaker. The Rainbow Group was engineered by Die Grünen to accommodate a far left Dutch group they had been instrumental in creating as well as a rag-tag of Italian communists and others without a home. It was only after the 1989 elections that the 'true' Greens were strong enough to form their own group and part from the Rainbows. See Sara Parkin's *Green Parties: an International Guide* (Heretic Books, London, 1989), for an absorbingly frank account of the Byzantine intricacy of European Green politics.
6. *Green Parties*, op. cit., p. 128.
7. Ibid., p. 123.
8. *Independent*, 27 July 1989.
9. Conversation with the author, 11 December 1989.

10. *Meet the Challenge, Make the Change. A new agenda for Britain.* Final report of Labour's Policy Review for the 1990s, October 1989.

CHAPTER ELEVEN

The campaigners

Margaret Thatcher is a rarity among national leaders in that she has a science background. While the greening of the electorate was rapidly taken aboard and she was impressed by the fact that acid rain could damage international relations as surely as it did stonework and trees, what caught her imagination were the profound implications of what was happening to the chemistry of the deeper atmosphere. Between the glimpses of the apocalypse when she spoke to the UN General Assembly in November 1989 on the unprecedented dangers of its own making which faced mankind was an appeal:

> it is no good squabbling over who is responsible or who should pay. Whole areas of our planet could be subject to drought and starvation if the pattern of rains and monsoons were to change as a result of the destruction of forests and the accumulation of greenhouse gases. We have to look forward, not backward. We shall only succeed in dealing with the problems through a vast international cooperative effort.

The challenge for the negotiators was as great as for any disarmament treaty. Time was too short to allow diversion into fruitless and divisive argument. There was a general acceptance within the green community that the speech was a good one, although it was grudging in some cases: 'Full of poetry and passion, but . . . short on policy,' wrote Tom Burke, Director of the Green Alliance.[1] What was needed was a lead not a lecture.

By the end of the year, Mrs Thatcher was in the chair of a new cabinet committee, Misc. 141, which had been given the task of preparing for the fast-approaching day when limits would be set on greenhouse gas emissions. The new policies which would emerge for discussion in an autumn 1990 white paper would take matter well beyond the December 1989 Green Bill, which does not include carbon dioxide among the gases for which standards are to be set. Cutting its emissions would go to the quick of industry, affecting road transport and energy output and involving potentially huge expenditures which the Treasury had, simultaneously with Misc. 141, set up its own committee to study. Both committees were the result of Chris Patten's desire to see more forward-thinking and more co-operation between departments.

It was a measure of how much the green political arena had changed that in December 1989 Jonathan Porritt, then Director of Friends of the Earth, was invited for talks at 10 Downing Street. Environmentalists were no longer 'the enemy within', as Mrs Thatcher had described them only three years earlier. The former eco-terrorists, feared for their readiness to ambush government whenever it marched into the green province, had suddenly become people worth informing and consulting at the highest level. They were almost – but not quite – on the same side. It could be argued, too, that they were beneficiaries of the heightened public awareness created by the increasing emphasis by government on the perils facing the environment. Many waverers must have reached for their chequebooks and signed on with a green campaign organization after hearing the latest pronouncement assuring them that there really was something to worry about.

There are dangers in becoming officially accepted. Complacency and a tendency to mute criticism of governments are among them. Becoming fashionable is another. The cutting-edge can become blunted. The campaign groups should ideally be voices calling from the wilderness on behalf of the wilderness. But then, of course, the wilderness grows less, or is saved (or perhaps tamed). The romantic days which have seen the bold David McTaggart striding the Pacific main in the Greenpeace ketch *Vega* to challenge the French bomb testers at Moruroa atoll may be passing simply because, in that particular case, the military reasons for testing are passing. The mood and issues of the 1990s are bound to be different from those of the 1980s. Climate change is already more important than acid rain, genetic engineering a more topical issue than saving whales. Changed issues do not necessarily throw organizations into an anxious state of transition; if they are any good, they are always in transition. That certainly applies to the three biggest international green campaign organizations: Greenpeace, Friends of the Earth and the World Wide Fund for Nature. They continue to grow in size and influence and they appear to be still some way from peaking. As they grow larger, they grow richer, they spend more on research and become more authoritative. Greenpeace was accused of scientific sloppiness, so it appointed a director of science and two people to work with him, a recognition in its way that zipping over the waves in pursuit of whalers was not enough to keep it in the forefront in these intellectually demanding days of climate change. But if you begin to get more solid and reliable – and perhaps institutional – do you also begin to get 'soft'? The authors of Greenpeace's official history, *The Greenpeace Story*,[2] go out of their way to refute the slur in their introduction:

Direct action remains the central theme of Greenpeace operations. This needs to be stated clearly because there is a current media cliché that Greenpeace is turning its back on such tactics and is becoming a more bureaucratic, softer version of its earlier radical self. This is demonstrably untrue; the number of direct actions continues in an upwards spiral. What is true is that, in recent years, such actions have been backed up by sophisticated political lobbying and scientific enquiry that have added strength to the organization's dramatic calls for change. Greenpeace's continued insistence on non-violent tactics, even when faced with violence, reflects both its cultural origins and its links with the other great movements for social change in the twentieth century.

Peter Melchett, Greenpeace's Executive Director in London, looked at the 'softness' issue from the aspect of whether being tough reduced influence.[3] 'People come to us and say, "Tone down the message and we'll talk to you", but we don't accept that. I don't think our direct actions are a problem reducing our effectiveness. Politicians may not talk to us publicly, but they do privately.'

The choice of heads for the campaigning organizations says a great deal about their memberships: solidly white middle-class. Lord Melchett, educated at Eton, former Labour Minister of State for Northern Ireland, became head of Greenpeace at the beginning of 1989, near the beginning of the membership boom. At Friends of the Earth there is (or was until mid-1990, when David Gee took over) his friend, the Hon. Jonathan Porritt, Old Etonian, son of Lord Porritt, former Governor of New Zealand. WWF's International President is the Duke of Edinburgh and its UK President is Princess Alexandra. The Duke has an old-fashioned huntin', shootin' and fishin' greenness which is a bit at odds with the panda-loving, badger-saving attitudes of most WWF members, but he is outspoken in his condemnation of 'the ruthless pursuit of economic development'. It is very English, of course, to fight for good causes under the cover of titled office-holders. Neither Melchett nor Porritt fits the usual patterns of lordship, but their presence is reassuring to members who like to see guarantees of moderation and respectability flown at the masthead.

A 'moral initiative' for nature

As the guardians and voice of nature the green pressure groups are more effective than any political party. The largest of them, Greenpeace, has, in the space of two decades, enlisted 3.5 million members worldwide and become rich enough to maintain a small fleet of boats and a research station in Antarctica. Its activities have led to one boat being badly damaged by a French warship, another rammed by the US Navy while trying to interrupt a missile test, and one sunk in harbour by French agents. In contrast to the other big pressure groups, Greenpeace does not accept funding from governments or firms. It has had, since its inception, an adversarial relationship with governments and its activists are not the sort of people with whom Conservative governments easily make friends, particularly when they block the outflow pipe of nuclear processing plants such as the one at Sellafield in Cumbria.

Greenpeace's cultural origins are Quaker. It was founded in 1970 by two American Quaker couples in Vancouver, British Colombia, who fell out with the Sierra Club, the long-established American conservationist organization, over its cautious stand on nuclear power and its refusal to become involved in big international issues. They recalled that in 1958 Quakers had made an unsuccessful attempt to interrupt atmospheric testing of H-bombs at Bikini atoll in the Pacific. The approaching US nuclear test at Amchitka in the Aleutians offered the chance of making a similar sea-borne protest. Confrontation with the United States was followed by the more dramatic confrontations with the French at Moruroa atoll where the goaded French beat up Greenpeace's skipper, McTaggart, rammed a protest boat and eventually murdered a photographer who went down with the sabotaged *Rainbow Warrior* in Auckland harbour in July 1985. The Pacific – and the oceans generally – became the main theatre of the movement's direct actions, although it has challenged French airfield builders in Antarctica and climbed towering chimney-stacks to plant banners condemning acid rain. Scenes of activists risking their lives to challenge lumbering factory ships generate publicity – and money – for causes of which the majority of people in Western countries approve.

In a sense Greenpeace volunteers are doing a lot of dirty work for the rest of us. How else would we have known about the slaughter of whales or seal cubs? How else could we have appreciated the dangers of ocean pollution, the extraordinary risks that are being taken with nuclear waste? The politicians will not tell us of what is happening to the environment. The vested interests of large corporations will not come clean unless we force them. That is why Greenpeace have had to take the moral initiative . . .[4]

'Moral initiative . . . a great movement for social change' (to quote *The Greenpeace Story* again)? One does not really think of saving dolphins as connected with social change, but essentially Greenpeace's claim is not exaggerated. In the West at least people have identified with a threatened world and the pressure groups have been the first instruments of that identification. Whales have been saved, so have seals, and nuclear power has had the question-mark over it enlarged. Greenpeace's international HQ produces an amazing amount of campaign literature: sea turtle fact sheets, *Everyone's Guide to Toxics in the Home*, five major arguments against kangaroo farming, ocean incineration of toxic waste, driftnets, the destruction of red coral, campaign to end nuclear weapon testing, save Antarctica, let's save the Mediterranean, and a great deal more.

Friends of the Earth comes from the same womb as Greenpeace, the Sierra Club. The period in which both were founded – 1969 in the case of the Friends, and 1970 in Greenpeace's – marked a transition from old-fashioned conservationism to a much more political and active defence of nature. The founder-president of FoE, David Brower, had resigned as executive director of the Sierra Club in 1969 after seventeen years in the post. His parting was a traumatic event which brough conservative lovers of scenic beauty and wide-open spaces into direct conflict with the new breed of activists, internationalist in their outlook. The Club had to decide, said Brower, whether to continue as a society of companions on the trail or take the position that 'the entire environment is the proper province of conservationists'. In a ballot, the Club membership decided to stay with the trails and Brower's suspension by the Board for 'dictatorial methods, dis-obedience and fiscal irresponsibility' was upheld.

In 1989, the membership of FoE in the United Kingdom leapt from 60,000 to 150,000, its income from £1,012,000 to

£2,144,000 and its expenditures from £914,000 to £1,996,000. Its advice was sought by all the parliamentary select committees involved in any way with the environment and it had a place on several of the government's consultative bodies. Internationally, it had observer status with the Food and Agriculture Organization, the International Maritime Organization, the London Dumping Convention, the International Whaling Commission, the Ramsar Convention on Wetlands and the International Tropical Timber Organization. There were branches in thirty-three countries. Despite its international image, FoE International's secretariat – currently in London – consists of one person. The work is done by the national branches. The British branch employs a fund raising team of eleven and an information staff of the same size. There are nineteen more employed on running campaigns ranging from tropical rainforests to air pollution and London's roads.

It is difficult to imagine the World Wide Fund for Nature encouraging activists to clamber up chimneystacks or challenge French warships. It raises funds for conservation and its concerns are with earth-bound pandas (it has helped set up ten panda reserves in China) rather than ocean-going whales. There is a note of disapproval in the response of an official to a question about relations with the other big campaigners: 'We work quite often with Friends of the Earth, less often with Greenpeace.' Royalty is in the van and the British headquarters are in Panda House, Weyside Park, Godalming, Surrey. The late Sir Peter Scott was the founder, in 1961 (WWF is the only one of the Big Three which has British origins), and international headquarters are in Gland, Switzerland, in the same building as the International Union for Conservation of Nature and Natural Resources (IUCN), with which it has close relations (and which, in turn, is close to UNEP). It bills itself as 'the largest private conservation organization in the world', with offices in twenty-three countries and more than one million members. In addition to the giant panda (its logo) it has 'reprieved' (WWF's word) the Indian tiger and the polar bear, the latter by persuading the five Arctic nations to restrict hunting. The African elephant was a notable beneficiary of WWF's activities in 1989, with two British lawyers working on its behalf during the negotiations on the Cites agreement on ivory.

A change of priorities

The world's great mammals have, however, been edged away from centre-stage recently. Since WWF's change to its present name from World Wildlife Fund it has, to quote a staffer in a rival pressure group, 'been trying to get away from its cuddly animal image'. WWF says the change was made 'to reflect more accurately the conservation work it does'. Nowadays, less than 10 per cent of its worldwide expenditure on conservation is directed solely on species protection. The priority has become saving habitats and solving environmental problems.

One of the projects by which it sets most store is the Plants Conservation Programme, run jointly with the IUCN. WWF found that finding a cuddly plant equivalent of the panda was difficult, so instead it focused on the value of plants to mankind, particularly medicinal plants. And at that point, of course, it swings into the issue of bio-diversity and the threat that destruction of the tropical rainforests, with their genetic richness, will mean more and more plants being lost for good. In Britain, too, there have been changes of emphasis. Until 1985 the bulk of its allocations to projects went on purchases of land for reserves. By 1989 reserves had slipped well down the list and education and public awareness had risen. The most significant change, though, was the appointment of a greenhouse gases and transport campaigner. Like the other campaign organizations, WWF had realized that the natural world can no longer be divided into isolated sectors of activity. All are united by the threat of climate change. Saving elephants, tigers and terns in the twentieth century will be a small gain if their habitats are ruined or destroyed along with man's in the twenty-first.

1989 was a marvellous year for the pressure groups. Even at its end the members were still rolling-in in their thousands – 4000–5000 a week in Greenpeace's case. Given the cycles in enthusiasm for green issues, maintaining the momentum may be a problem, but the example of West Germany is encouraging. Its population is approximately the same as Britain's but its Greenpeace has 600,000 members, double the British membership. Looked at collectively, the green movements in Britain are already numerically impressive. To coincide with World Environment Day on 5 June 1989, *The Times* published a survey which showed that the membership of fifteen green organizations was climbing towards parity with the membership of unions affiliated to the TUC. If the current growth rates continued, it would be close to 6 million by 1992, with the TUC at 8.8 million and declining. It was an unfair comparison, of course. Most of the 2 million who belong to the National Trust (one of the fifteen) do so for the sake of visiting the houses it preserves, not because it is green. All the same, the growth rate of this group, all broadly linked by an interest in the countryside and the environment, was evidence of the power building up behind the green movement in the 1980s and 1990s.

There is an obvious danger of excessive duplication when broadly similar organizations conduct broadly similar campaigns. The rainforests are an obvious example. As fundraisers they are unparalleled. The urban middle-class heart goes out to jaguars, exotic Indians and magnificent trees with a generosity that is not evident in the cases of kangaroos and red coral. FoE is heavily involved in the rainforests; WWF considers them its 'major international priority'; Greenpeace started its rainforest campaign in 1989 and pondered the possibility of seaborne harassment of ships carrying timber exports. 'We are in competition, and I think that is good, but there is a lot of information-sharing,' said Peter Melchett. 'There are informal networks through which we let one another know what we are doing and we meet and talk. For instance, Greenpeace and Friends of the Earth had ten people each at a meeting recently (towards the end of 1989) to talk about the Green Bill and other issues.'

Ad hoc alliances of green organizations take place from time to time, and at the end of 1988 Greenpeace, Friends of the Earth and the World Wide Fund for Nature jointly threw down a 'green gauntlet' to the government, urging it to turn propaganda about the environment into action. A list of thirty measures ranging from recycling resources to doubling overseas aid were proposed. It made very little impact. They appeared to be of one mind again when they appeared jointly to give evidence before the Commons' Energy Committee on the greenhouse effect's implications for energy policy. This time, though, there was an interesting addition to their ranks, the Association for the Conservation of Energy, a combination of think-tank and pressure group mainly financed by manufacturers of equipment for conserving heat and energy. 'We will have to tackle the major issues,' Peter Melchett had said, a shade regretfully, when asked how Greenpeace, with its policy of high-profile direct actions, would deal with the nebulous elements of climate warming. They sent along their air pollution campaigner to the energy committee hearing. But here was the future. The grey enigmatic god carbon dioxide was taking over from Pan.

Notes

1. *Independent*, 10 November 1989.
2. Michael Brown and John May, *The Greenpeace Story* (Greenpeace Books and Dorling Kindersley, London, 1989).
3. Interview with author, 19 December 1989.
4. Fund-raising letter, 1989.

CHAPTER TWELVE

The uncertain politics of carbon dioxide

There is indeed something godlike about carbon, its omni-presence in all things living and dead, the grand cycles of time that move it in its vehicular form carbon dioxide between land, sea and air, the fertility it provides and its release by fire. It manifests itself in diamonds as well as in coal, and it has more compounds than any other element. A branch of science, organic chemistry, is witness to their complexity. In combination with oxygen it forms the gas carbon dioxide. A colourless gas with a faint tang, it can produce narcosis and even (after a lengthy exposure at a concentration of 5 per cent or more) unconsciousness and death, but it is as a greenhouse gas emitted in the course of producing man-made energy that it has become central to our end-of-century concerns.

That sort of centrality does not automatically make it an easy issue for national politicians, though. All profess greenness nowadays, of course, but climate change has not yet become part of their instinctual combat kit with its codes telling them instantly what to do when action is joined over, for example, trade union legislation and welfare payments. The first Commons debate on world climate change came only two days after the prime minister's theme-setting UN speech on the environment in November 1989, yet at no stage in its five-hour course was it attended by more than fourteen MPs. The debate was opened by David Trippier, Minister of State for the Environment, who said the House could debate no issue of more immediate importance . . . and then announced that he would have to leave early to present the Young Environmentalist of the Year award at a lunch.

Attitudes on an issue which hardly ever figures in the popular press and does not fit into a ready-made political rut are bound to change slowly. The warning signs were, in any case, hoisted slowly and cautiously by the scientists. Early in 1980, Whitehall's newly-created Interdepartmental Group on Climate Change published a report[1] on the implications of its subject for the United Kingdom. The Thatcher administration was still new to the power it was to hold for the rest of the decade and a reader might reasonably expect to find something seminal in the report. But its reluctant, even remote, tone is that of bureaucrats anxious to distance themselves from what might prove to be no more than fashionable scientific nonsense. 'Some people' had argued that increasing atmospheric carbon dioxide had serious implications for the climate, but the committee had not come to any conclusions on that. Recent extremes of weather had given 'speculations' credibility with the public, but the Meteorological Office scientists took the view that the variations in weather were compatible with established climate patterns. The scientists were being cautious and so were the civil servants; this meant that British politicians could happily continue to struggle over familiar issues until 1988 when a devastating drought enabled scientists in the USA to get the ear of the Senate and pour into it a story sufficiently sensational to alarm the rest of the Western world.

The scientists who man the climatological observation posts are less cautious than they used to be now that the general theory of climate change has become common property; government servants are more confident and outspoken; and, while scepticism remains, no cabinet minister is likely to denounce the theory and the accompanying evidence as hysterical. In 1978–9 the British government spent a modest £250,000 on research, including the development of climate models. By 1989–90 the budget was £15.54 million (0.53 per cent of the total government expenditure on civil science and technology research and development), with an additional £760,000 over two years for the Intergovernmental Panel on Climate Change and £5 million pledged to fund the new centre for the prediction of climate change announced by Mrs Thatcher in her UN speech on the environment.

The valedictory thoughts of Sir Ian

The moving spirit behind the November 1989 Commons debate was Sir Ian Lloyd, who delivered a grand and gloomy swan song as chairman of the Commons' Energy Committee. He was, he said, one of that small group of political eccentrics who browsed around the dustbins of policy hoping occasionally to come across the word 'science':

> I make a more general criticism. It is not so much a failure of one or other particular party, it is a failure of our culture to recognize the significance and importance of science and to elevate it within the nation more continuously and more substantially than governments of either party have chosen to do during the past 25 years.[2]

Sir Ian is a former president of the Cambridge Union, whose career has been woven out of a mixture of Conservative politics, economics and technology. He has tried unsuccessfully to persuade government that it should devote a fixed proportion of either GDP or the defence budget (which has some logic as an idea, since the Meteorological Office is funded through the Ministry of Defence) to research into global warming; and failure in that respect has reinforced his view that governments across the world find inexpensive words preferable to expensive actions. He continued:

> Although the political will has been expressed in vigorous, interesting and dramatic terms at international conferences, in my judgement the political will to follow through that policy right across the globe to a degree that will be effective does not exist. That political will does not exist because public opinion has not yet begun to be aware of the scale of possible changes that governments will ask their publics to accept. [And even if the political

will did exist, he doubted whether the institutions which
would have to translate it into collective action were
strong enough to do so.] Our international institutions
are frail, fragile and tend to be powerless. We are
moving into an era when frailty, fragility and powerless-
ness are simply not enough. The proper institutions must
be created.

What sort of 'proper institutions' did he mean? 'I was
thinking of a general problem,' he said later,[3]

> which is how far parliaments should surrender part or the
> whole of their responsibilities. It is my belief that the
> nature and scale of the change needed is unprecedented.
> I just don't believe that a conglomeration of nation-states
> will be able to deal with it. We are facing a situation
> where the human race will have to govern itself. No
> existing institution has the power to encapsulate and
> carry out policies, so we will need new ones. At the
> moment the world is too fragmented to deal with the
> crises that will emerge.

International conventions were all very well, but they were far
too easy to breach. War as the ultimate sanction was not a
credible solution. Every instinct and all the forces in political
democracy were against the surrender of sovereignty, but we
had to move down the road towards accepting it. Governments would face the prospect of catastrophic events and
massive expenditures. They would have to make political
judgements on how far to reduce standards of living.

Sir Ian thinks further ahead than most governments care to
go. Surrendering sovereignty to a UN body or its equivalent is
not on even the sketchiest draft agenda for the future. 'Ian
Lloyd is terribly pessimistic,' a lobbyist in the energy field told
the author. 'He doesn't think the world will be able to agree
on what should be done on global warming. He may be right.
There are too many vested interests.'

Vested interests in sovereignty, in industry, in national
wealth, in political power. Whichever way you turn in trying
to halt the elements of global warming, the costs mount; and

as they mount, so do taxes and the perils of inflation. The Conservative government in Britain in 1990 still gave priority to fighting inflation, which was rising more steeply than in its commercial competitors. The promises to cut taxes had been brushed to one side, but it was unlikely that the government would wish to fight the next election against a background of undertakings that would entail their actually being increased. The government's position is that the major decisions on countering global warming will have to wait on international agreement at the 1992 UN conference on the environment, which would project them beyond the present government. Unilateral action is out, since its cost would reduce industry's competitiveness in international markets. So no gestures. It remains, too, the official conviction of the government that market forces rather than regulation will create the necessary responses to global warming.

Limiting carbon dioxide emissions depended not on regulation but on the aggregate decisions of millions of individuals remote from government, John Wakeham, the British Energy Secretary, told the World Energy Conference in Montreal in September 1989. There were major drawbacks in concentrating solely on a 'regulatory route'. By contrast, there were clear advantages if people were motivated by their own self-interest to restrain greenhouse emissions. A tax or taxes on energy generally were possible approaches. Moreover, countries should price fuels to reflect their full costs, including environmental costs. They needed to develop market mechanisms which would take account of the cost of environmental effects, and to develop the principle of the 'polluter pays'.

'Market forces' is regarded by some as a doctrinal slogan as nebulous as carbon dioxide itself. 'The cost of polluting the environment is not recognized by the market mechanism,' the Institution of Electrical Engineers remarked drily in its written evidence to the House Energy Committee hearings[4] on the greenhouse effect and its implications for energy policy. There was no way at present of costing one environmental 'pollutant' against another. 'How many times worse than 200 unsightly windmills is the radioactive product of a nuclear power station or the CO_2 emissions of one coal-fired station?' asked the Energy Committee in its report of the hearings.

Counting the cost of carbon

It is estimated that every year more than 200 billion tonnes of carbon are exchanged between the atmosphere, living things and the ocean in a natural cycle. To that, humanity adds about 5.6 billion tonnes from burning fossil fuels, of which the developed nations are responsible for 70–75 per cent, with power stations the worst offenders. In the last two decades world production of electricity has roughly doubled, with the developing nations pulling towards overtaking the developed. The best guess prediction is that by 2030 the concentration of greenhouse gases of all kinds will be equivalent to a doubling of carbon dioxide (if CO_2 was the only gas, the doubling would not take place until 2080). The consensus is that stabilizing the climate (i.e. halting the warming) will require a 50 per cent cut in carbon dioxide emissions.

The Brundtland Commission sketched two scenarios of energy consumption in *Our Common Future*, one high, the other low. It took as a starting point a figure of about 10 Terrawatts (one Terrawatt equals one billion kilowatts) for global energy consumption in 1980. If per capita consumption remained static, an increased world population would still need 40 per cent more energy by 2025. But if energy consumption became uniform, with people in the developing world using as much per capita as those in the industrial world, global consumption would rise to 55 Terrawatts (TW). Even if that were scaled down to 35 TW, the world in 2030 would still require 1.6 times as much oil as in 1980, 3.4 times as much natural gas and nearly 6 times as much coal, and nuclear capacity would have to be 30 times greater than in 1980. The low 11.2 TW scenario would require the industrialized nations to halve their energy consumption by 2020, while the developing world would more than double theirs. One can, of course, go on writing scenarios and juggling the figures almost indefinitely; the only sure thing that emerges is that stabilizing the climate by halving CO_2 emissions is going to demand a very great – and expensive – effort by the industrialized nations. It may well prove to be an impossible task.

Costings of domestic expenditures in industrial countries on reducing emissions tend to brush away the minor irrelevancies of the millions and concentrate on refining estimates to the nearest billion. The Electricity Council told the Energy Committee that it would be possible to meet the Toronto Conference target of a 20 per cent reduction in carbon dioxide emissions by 2005. The biggest item would be thermal insulation – £5 billion. Other measures would involve more combined heat and power (CHP) stations, which, instead of discharging into the atmosphere the 60–70 per cent of the primary energy input which ends up as waste heat, would use it to provide hot water and heating in homes and commercial and industrial buildings; and, of course, a switch from coal – which produces more carbon dioxide per ton than any other fuel – to nuclear, gas, oil and 'renewables' such as windfalls and tidal barrages. The council was doubtful whether the public would want to pay for the changes. The Central Electricity Generating Board estimated that achieving a 20 per cent cut through switching from coal to nuclear, gas and oil would require an additional capital expenditure of £12 billion. The effect on electricity prices would be 'very severe'. Dr Ken Currie, the head of the government's Energy Technology Support Unit, had similar reservations when he told the Cabinet at a seminar on climate change in April 1989 that technically it was possible to halve carbon dioxide emissions (the longer-term Toronto objective). On the other hand, there were 'practical obstacles and controversial issues' to be overcome. 'To put the economic aspect into perspective, it would need at least a doubling of the price of energy to make some of the supply options attractive to investors.'

There are broadly three ways of reducing mankind's output of carbon dioxide. Power stations can be made cleaner and more efficient, so that they use less fossil fuel, particularly high-carbon coal; energy can be used more economically by the consumer; and there is nuclear energy (the subject of the following chapter), which whatever else it is capable of doing, does not produce carbon dioxide. To take more efficient power stations first; there are advanced technologies using fossil fuels (gas turbines, in particular; natural gas produces less carbon dioxide than coal and oil) which are considerably more efficient and produce less carbon dioxide. Under the heading of 'renewables' are windmills, tidal barrages, wave power, solar energy and the 'hot dry rocks' of geothermal power. Windmills and tidal barrages (and perhaps wave power) are regarded as having promise even if the former do spoil landscapes and the latter ruin the mudflats on which ducks and wading birds feed. However, estimates of the space needed for a battery of windmills capable of producing the same amount of electricity as a normal power station, fossil fuel or nuclear, range from 100 to 200 square miles. Sunlight converted into useable energy by photovoltaic cells is of economic value only in countries with a lot of sunshine. Geothermal 'hot dry rocks' are not strictly 'renewable' since the rocks, deep within the earth's crust, lose their heat when it is tapped. Their potential is considerable, though, and the Department of Energy estimated that they could provide 10 per cent of Britain's annual energy requirements over 100 years.[5] Methane gas from rubbish tips is another potentially valuable source of energy.

Just as you can remove sulphur from power station emissions so you can do the same with carbon dioxide. Theoretically that is. Studies carried out by the Brookhaven National Laboratory in the United States suggested that 88 per cent of the carbon dioxide could be removed by an absorption-stripping system. However, carbon produces more then 3.5 times its own weight in carbon dioxide[6] so there is a problem in disposing of it. The most feasible means is to pump it as a gas or liquid through a pipeline to the ocean, where at 500 metres, most of it would stay out of contact with the atmosphere for many years, since liquid carbon dioxide has a density greater than seawater. Even safer – but more costly – is to pump it to a depth of 3000 metres. However, nothing is known about the long-term environmental consequences – or even effectiveness – of using the ocean beds as dumps. Other possible places for disposal are abandoned oil and gas reservoirs and aquifers. Equipping a plant to cope with carbon dioxide is likely to double its cost. Add to that the 17–20 per cent of the power output consumed in stripping the carbon dioxide and the outcome is probably a doubling in the cost of the electricity produced. It is a sign of the times that despite the cost research is continuing.

Leaving out nuclear power and the renewables, the most favoured low carbon dioxide fuel is natural gas (which is largely methane, another greenhouse gas). Emissions from a gas-fired plant are about half those from a coal-fired one. However, there is a snag. Natural gas supplies will begin running down in the twenty-first century. The United Kingdom's reserves are likely to be empty in forty years even at the current rate of consumption. Worldwide gas, and oil, prices are expected to start rising early in the next century. Moreover, the use of gas is not likely to have much impact globally. The Energy Committee's own calculation was that it would require a 90 per cent increase in gas consumption worldwide to reduce carbon dioxide emissions by 10 per cent.

The double-glazed future

One comes back to what most people are agreed on: in the short term the simplest, cheapest, quickest and most effective means of reducing carbon dioxide emissions is to use energy more efficiently, in both homes and factories. Most developed countries are doing so already – and in fact have been doing so for the past sixty years. Energy saving does not date from the first oil crisis. Britain is not in the first rank of energy-efficient nations, but, taking 1975 as 100, the energy requirements backing each unit of GDP fell from 115 in 1970 to 81 in 1987. That was better than the OECD (Organization for Economic Cooperation and Development) European average of 106 falling to 89 over the same period.[7] Britain's annual energy needs were almost unchanged between the beginning and end of that period, but real GDP rose by 42 per cent. It was Japan, the most energy-intensive country in the world, though, that showed what can be done. Between 1973 and 1985 GDP there rose by 46 per cent while energy use fell by 6 per cent, and that, according to the Association for the Conservation of Energy,[8] put the United States 'at a $200 billion disadvantage to Japan as a result of a poorer energy efficiency performance'.

Look at the figures for electricity generated (and that is where most of the carbon dioxide comes from), however, and a rather different picture emerges. While the direct use of fuels for purposes other than electricity generation has gone down, the output of electricity has gone up. In fact, it almost doubled in the OECD between 1970 and 1987 and the trend remains upward. Coal-fired power stations were responsible for more than 41 per cent of the electricity in 1987, with nuclear power second with 22 per cent and gas third at 9 per cent. So it is possible to have more efficient use of energy but an increase in output from the sector which produces the most carbon dioxide.

The Department of Energy sprang a surprise in the autumn of 1989 when the paper it produced for the 'responses' group of the Intergovernmental Panel on Climate Change[9] suggested that, with more efficient use of energy, Britain could cut its consumption (and hence carbon dioxide) by 60 per cent by 2005 or 2020. A more realistic figure, it admitted, was 20 per cent. The government acknowledged the need for a target in May 1990 by undertaking to stabilize carbon dioxide emissions at the 1990 level by 2005. Coincidentally with the DoE report, Friends of the Earth produced a non-nuclear package of measures[10] which, it claimed, would produce a 46.5 per cent reduction in emissions in all sectors other than transport, which it excluded. What is striking in any analysis of this sort is how much carbon dioxide is attributable to consumption of energy by the domestic sector: 29.1 per cent in 1987 as a result of all those central heating systems, television sets and cooking stoves. Industry was only slightly ahead at 29.5 per cent, and the transport sector – the fastest growing source of carbon dioxide – was third with 23.5 per cent.

FoE made a number of recommendations on taxes, incentives and greater spending on research (including the government's Energy Efficiency Office, which, remarkably, given the nature of the times, saw its budget restricted to £15 million a year for two years in succession – a loss in real terms). The emphasis generally is on the carrot of incentive rather than the stick of taxes and sharply raised prices. The latter were 'a blunt instrument', said Stewart Boyle, of the Association for the Conservation of Energy, in his address in December 1989 to Chatham House's Fourth International Energy Conference. 'Very high increases could lead to a recession as happened in the late 1970s and early 80s due to the oil price hikes. This would tend to slow down capital turnover of old, inefficient equipment, exactly the opposite effect that is required.' He quoted a study of all fuels by the Henley Forecasting Centre which concluded that doubling the fuel costs over five years would simply keep energy consumption level through the 1990s. As for cars, a study[11] of twenty-five countries concluded that a 10 per cent fuel price increase produced only a 3 per cent improvement in fuel economy. Price rises were ineffective unless they were so swingeing as to be ruinous.

A rule of thumb is that investment in energy efficiency is expected to pay for itself in two years. If investments have already been made in recent years, the average investor will not bother to make further improvements until energy prices begin to hurt. A visitor to ACE's premises noticed that the windows in the director's office were not double-glazed. The director, Andrew Warren, admitted that even in that citadel of energy conservation he had come to the conclusion that the savings would not justify the cost. Mr Warren's equivalent in Japan would almost certainly have had his windows double-glazed. Japan has few energy resources of its own and it has learnt to be frugal: energy prices are approximately twice those in Britain.[12] It is also the world's most dynamic and successful economy. Measured on the carbon dioxide scale, it emitted 2.109 metric tons per capita in 1986 from fossil fuels and cement production;[13] Britain in the same year emitted 2.938 metric tons per capita from the same resources; the United States 5.005; France (which relies on nuclear power for 70 per cent of its electricity generation) 1.794, and India 0.187.[14]

'Energy demand is shifting, from production to pleasure,' declared Lee Schipper and Andrea Ketoff,[15] at the December 1989 Chatham House conference. They were talking of the developed countries, of course, perhaps with North America mainly in mind. Consumers had always wanted bigger cars, holidays in places that were further away from home, bigger refrigerators, larger homes and well-lit amusement parks. They could realize those ambitions because: 'Low fuel and electricity prices make these goals that much easier for

consumers, who are not worrying about the energy implications of their choices: the colour of the refrigerator becomes
more important than its energy use.'

Since 1972 (the run-up to the Middle East oil crisis), the
amount of heated space per capita in homes in most OECD
countries had grown by 25 per cent. The number of cars and
total miles driven was at least 25 per cent higher in most large
countries. Air travel was growing rapidly, while the role of
mass transit on the ground was shrinking almost everywhere.
A combination of cars and aircraft provided over 80 per cent
of passenger transport in Western Europe and over 95 per
cent in the United States. Between 1972 and 1985 there had
been significant energy savings in OECD countries. The
extent to which the momentum towards more efficient use of
energy for heat had slackened since 1985 had surprised them.
Much the same applied to new refrigerators and electrical
appliances. Consumers were not choosing efficiency and
manufacturers were not improving efficiency as much as they
had in the past. The results suggested a plateau of energy
efficiency had been reached. Manufacturers of energy-using
equipment for aircraft engines to cars were aware of consumer
indifference. Low energy prices were a problem. 'Since most
experts seem to preclude a major upturn in energy prices
before the end of the century, we may have to develop a
combination of policies and taxes to stimulate the market for
improvements in efficiency *now* as insurance against disrup-
tions from climate change *later*.'

The developed North worries about how to make luxury
more responsible with better light bulbs, the impoverished
South struggles to put light bulbs into its shanty towns.
Efficiency plateaux are still a long way from becoming a
feature of Southern energy economics. The scale of the divide
between North and South was sketched in a paper read at the
December 1989 Chatham House conference by Rajendra
Pachauri, Director of the Tata Energy Research Institute in
New Delhi.[16] The North's share of the world's energy
consumption was expected to decrease from around 70 per
cent to 60 per cent by 2020, but in that time the proportion of
the global population living in the South would have grown
from 75 per cent to 80 per cent. Even with an allowance for

higher economic growth rates in the South than in the North, the gap in per capita consumption levels was likely to grow. Improved standards of living in the South would depend on increases in consumption of energy, but there was small prospect of its being consumed as efficiently as in the North. At best, efficiency improvements in the near future would shave off two or three percentage points in the rate of growth in energy demand. It was unlikely that countries like Brazil, China and India, which would be the largest consumers of energy in the future, would succumb to international treaties or conventions limiting energy use at the cost of economic development.

The question needs to be asked, whether such limitations would be valid on moral, ethical or social grounds given the disparities that exist in living standards . . . . The cost of limiting emissions of greenhouse gases has to be borne by those societies which are responsible for the cumulative increase in their concentration levels.

The North can argue, of course, that in a world of finite fossil fuel resources inefficiency in itself is immoral. Carbon dioxide emissions per head of population may be low in the developing countries, but they are high when measured against units of GDP. But the paraphernalia of better light bulbs, lean-burn engines and cavity-wall insulation with which the North supports its arguments are likely to be dismissed as irrelevant in parts of the world which for the most part have no need for cavity walls. The South's case is that for the sake of the ever-improving standards of life which the North's electorates demand, it squanders resources, poisons the atmosphere and then, to stem the environmental degradation it has caused, demands that the impoverished South forgo equality with the North in the name of salvation!

Shares in the greenhouse

By the end of the 1980s all the major governments in the West were carrying out reviews and costing of environmental policies. It is a virtual certainty that the developed world will have to set standards for itself, probably through commitments within the ECE or OECD framework, before attempting the infinitely difficult problem of persuading the South to join it in global agreements.

One means of persuasion which has been suggested is the creation of 'greenhouse gas shares', a virtual currency consisting of marketable carbon emission permits. Something similar in principle has been tried in the United States, with the issue of tradable acid pollution permits. It sounds at first a morally dubious proposition, but Michael Grubb argues in *The Greenhouse Effect*[17] that it is more promising, and fairer, that any other scheme for international control.

The principle which would have to be established first of all is that each person in the world (all those aged 21 or over, suggests Dr Grubb) would have an equal share in man-made carbon dioxide emissions, regardless of whether he lived in a rich developed country or a poor one which produced hardly any carbon dioxide at all. The next step would be the international negotiation of a target for worldwide emissions over a set period, perhaps a decade or two. The permits would then be allocated by an international agency to governments. Countries whose emissions exceeded their allocations would be obliged to buy (or perhaps lease) permits from countries whose output of carbon dioxide fell below their allowance. The permits could be traded for what the seller wanted – technology, energy-efficiency equipment, development programmes, and so on. In that way, the rich polluters would be obliged to pay for their excesses through a transfer of technology and wealth to the poor, who in time would come under pressure themselves not to allow their emissions to reach levels which would force them to buy permits.

A market-based scheme of that sort has its attractions; and it has its problems, too. What would be the value of the 'currency'? If the value was fixed by bidding, would it mean that the richest nations cornered what they needed and left the not-so-rich scrabbling for the left-overs? Would big but impoverished polluters like the Soviet Union and the East European countries accept it? How would it be monitored and regulated to ensure that those who cheated were penalized?

But even if it is not exactly this scheme, it is certain that any global agreement will have to be supported by a mechanism which allows the developing countries to pollute more while they catch up with developed countries which are simultaneously reducing their emissions. Rich countries are not going to accept proposals from India and the Group of 77 countries for an environmental aid fund which would be administered by the developing nations, but transfers of technology and financial aid will be part of the eventual package. Rajiv Gandhi's call when still India's prime minister at the September 1989 Non-Aligned summit for an $18 million fund based on contributions equivalent to 0.1 per cent of GDP by UN members has the support in principle of the influential Mrs Gro Brundtland of Norway. The argument will be harsh, and if the threat of disaster is vivid enough to bring about agreement, it may be that nations will find that, in the interest of common purpose, they are little by little surrendering the sovereignty of which Sir Ian Lloyd spoke.

Notes

1. *Climate Change: its potential effects on the United Kingdom and its implications for research* (Cabinet Office, HMSO, 1980).
2. *Hansard*, no. 1499, 6–10 November 1989.
3. Interview with author, 13 December 1989.
4. See Sixth Report of the Energy Committee, *Energy Policy Implications of the Greenhouse Effect* (House of Commons, HMSO, July 1989), vols 1–3.
5. UK country study for the Intergovernmental Panel on Climate Change, Response Strategies working group, Energy and Industries sub-group.
6. This seeming mystery is explained by adding up the components of a molecule of carbon dioxide. Carbon has an atomic weight of 12, oxygen (with which it combines to make carbon dioxide) of 16. The molecule is composed of one carbon atom and two oxygen atoms. Added together they make 44, or 3.7 times the weight of the original carbon atom.
7. *OECD Environmental Data; Compendium 1989* (OECD, Paris); see section 12, part II.

8. Memorandum submitted to the House Energy Committe, February 1989.
9. See 5 above.
10. *Getting out of the Greenhouse: an agenda for U.K. action on energy policy* (Friends of the Earth, London, 1989).
11. D.R. Bohi, *Analysing Demand Behaviour* (Johns Hopkins University Press, Baltimore, 1981).
12. Cecil Parkinson, then Energy Secretary, Q.510, Vol. III. Sixth Report of Energy Committee, op. cit.
13. A minor contributor, but the only industrial source of carbon dioxide other than fossil fuels. The calcium carbonate in the raw material (usually fossil rock, such as limestone) is broken down in the cement-kiln into carbon dioxide and calcium oxide. In the process the carbon dioxide is released into the atmosphere.
14. *Estimates of CO_2 Burning* (Oak Ridge National Laboratory, Tenn., 1989, Environmental Sciences Division. Publication no. 3176).
15. Lee Schipper and Andrea Ketoff, *Energy Efficiency: the Perils of a Plateau* (International Energy Studies, Lawrence Berkeley Laboratory). [Paper read at December 1989 Chatham House Conference.]
16. Rajendra K. Pachauri, *Energy Efficiency in Developing Countries: Policy Options and the Poverty Dilemma.* [Paper read at December 1989 Chatham House Conference.]
17. Michael Grubb, *The Greenhouse Effect: Negotiating Targets* (Institute of International Affairs, London, 1989).

CHAPTER THIRTEEN

Nuclear power: carbon-free but costly and feared

It was mid-January 1990 and the members of the Commons' Energy Committee were plodding towards the ever-receding horizons that contained the answer to their question: was it worth continuing to fund the fast breeder reactor (FBR)? The representatives of the UK Atomic Energy Authority who had been summoned to attend the session were zealously optimistic about the reactor's prospects, but defending the future of the nuclear industry at the beginning of the 1990s was no easy task. Uranium prices had slumped. New reactor orders were virtually non-existent. Only resource-starved, energy efficient Japan, the first and so far only victim of nuclear warfare, carried the torch of nuclear power with undiminished enthusiasm. In Britain, two months earlier, the government had decided in a last-minute about-turn that its nuclear power stations, once the shining lamps that would light the way to the sustainably developed future, were unsaleable and so would be withdrawn from privatization of the electricity industry. Was it the death knell of the nuclear industry or merely curfew time? Whichever bell was tolling, the question-and-answer on the specific issue of the FBR revealed that the committee was dealing with a legendary creature indifferent to the usual processes of judgement, time and fate. Never before has a machine been allowed such a long and cosseted lead-time. Untold billions have been poured into research and building working reactors in return for a relative few penn'orths of electricity. Research on the fast breeder reactor started in the United States in 1945 and the answers extracted by the committee indicated that the product was unlikely to be needed until practically the middle of the twenty-first century, by which time uranium resources would be running low and prices accordingly high. Or so it was speculated. A grand dismissal of the niggling demands of penny-pinchers who wanted results tomorrow came from the chief witness, John Collier, Chairman of the UKAEA and head of the newly-formed Nuclear Electric utility. When pondering the future of the FBR one needed time-frames of 100–200 years, he declared. No, it was not possible to say when a commercially viable FBR would be built.

The expensive (and dangerous) charm of the FBR derives from its ability to breed fissile plutonium 239 for use as its fuel from uranium's most common isotope, non-fissionable U238. This is over 100 times more abundant than the fissile U235 which fuels conventional thermal reactors like the pressurized water reactor (PWR). However, some breeding takes place in thermal reactors, so the actual gain is that a fast reactor can generate about sixty times as much electricity as a thermal reactor from each tonne of uranium. Since fast reactors can breed from the depleted uranium in the spent fuel from thermal reactors, it can be calculated that in FBR terms British stocks of depleted uranium are comparable to the country's coal reserves. On a global scale, the same calculation doubles world fuel resources. No one loves plutonium, but equally a tamed carbon-free and virtually inexhaustible fuel has its attractions in a world in which increasing demands are made on finite energy resources. Electricity generation would no longer be the major contributor to the greenhouse effect and its fuel would be 'sustainable'.

In the late 1940s and 1950s the fast reactor was held in such esteem that other forms of nuclear power generation were seen as filling in time until the FBRs were ready to go on stream. The experimental model at Dounreay, in Scotland, was the first in the world, in 1959, to become operational. In the same year it was considered likely that a commercial FBR would be ready in 1965. That forecast later slipped back to 1970. In 1967, Sir William Penney, then chairman of UKAEA, thought the first commercial reactor would be in operation in 1979. In 1975 Dr (later Lord) William Marshall, the country's leading champion of nuclear power, lamented before a House of Lords' committee that at the current rate of progress there might be only two fast reactors on line by 2000.[1]

And so it has gone on, although by the end of the 1980s it was evident that the patience of even a government as dedicated to the eventual triumph of nuclear power as Mrs Thatcher's was showing signs of severe strain. After a review of the state of research, the government decided, in 1988, that it could not justify continued expenditure of £100 million a year on a project with such an elusive prospect of commercial return. The prototype fast reactor at Dounreay would be funded until 1994 and its reprocessing plant for a further three years to allow it to deal with spent fuel. From 1990 there would be a core research and development programme costing £10 million a year, just enough, in the government's view, to allow continued collaboration with the French and Germans on the European Fast Reactor (EFR).

The agreements on EFR had been signed in early 1989 and in October of the same year outline permission was given for the building of a £300 million 'demonstration' reprocessing plant at Dounreay to recycle EFR's spent plutonium fuel. The money was as shadowily outlined as the permission, but, like the EFR agreement itself, it revealed the government's unwillingness to relinquish once and for all its place in the search for inexhaustible energy.

But what is enough in the voracious world of nuclear finances? Not what the government was offering, said the UKAEA in its written testimony to the Commons' Energy Committee.[2] The British commitment to EFR was being undermined and Britain placed in a 'weak position' relative to its partners. 'Outstanding progress' had been made with EFR's design and it was planned to start construction in 1997 and have it ready for operation in 2005. After a minimum five years' operating experience it would be time to think about taking a few orders for commercial stations, the first of which should come on stream about 2020 – always provided, of course, that the cost and dwindling resources of uranium made it commercially worthwhile:

The strategic need for diversity and ultimately for the

replacement of fossil fuels has been the driving force for the development of nuclear power in the U.K. In turn – since uranium reserves are finite – the efficient exploitation of nuclear power itself ultimately depends on the fast reactor. Since the 1950s this has been the nub of the strategic and commercial argument for developing fast reactors, and it has not changed. Recently, however, it has been given added emphasis and urgency by the growing appreciation that the damage being done to the environment by the use of fossil fuels may require drastic restrictions on their use. We therefore believe that greater reliance on the fast reactor is a necessary consequence of such restrictions and that the timescale for the large scale introduction of fast reactors may need to be brought forward.

That is the classic case for the FBR. The EC Commission said much the same in its submission to the Energy Committee: 'At the present time the FBR is the only reactor type which could, if introduced early enough, extend the lifetime of our uranium resources to the end of the next century – and beyond.'

Strategic needs do not, however, dominate human (or even commercial) reactions. No branch of nuclear technology is more detested and feared by the anti-nuclear and green lobbies than fast breeders. Opposition to fast breeders in Europe has been vigorous. One man died in the 1977 protest at the Creys-Malville FBR site in France: the Kalkar fast reactor in West Germany is mothballed (at a reputed cost of £30 million a year) largely as a result of green agitation. In a memorandum written for the Commons' Energy Committee, Andrew Holmes, the editor of *Power in Europe*, wrote, 'In the course of my work as a journalist, I have met a few environmentalists who can envisage some kind of long-term accommodation with the nuclear power industry, if waste disposal and other problems are satisfactorily solved; I have never met an environmentalist who was not wholly and implacably opposed to the FBR.'

FBRs mean plutonium and plutonium means warheads and, potentially, nuclear holocausts. If fast breeders proliferate how can the spread of plutonium to countries which would have no scruples about making their own nuclear weapons be checked? One measure of how the dangers of plutonium are assessed is the Japanese plan to build an £88 million coastguard vessel to escort shipments of plutonium oxide from the Sellafield reprocessing plant in Cumbria for use in its prototype fast reactor at Monju. Air transport was ruled out because of the danger of crashes on land, which in this case, because of the route, might turn out to be US soil in Alaska. Sending it by sea, on the other hand, would run the risk of hijackings by terrorists or even governments. A ship would have to be guarded by a warship, the Americans, who supplied the original plutonium, insisted.

As for commercial reactions, they can be judged by the fact that the capital cost of a first FBR is estimated by the British nuclear industry at between 20 and 30 per cent greater than that of a PWR, and generating costs 20 per cent more. Even the cost of generating electricity from the technically success-ful French Superphénix is twice that of a PWR.

Unloved and out in the cold

If uranium prices are going to swing in favour of fast reactors, they have a long way to go. At the beginning of the 1990s the spot price for uranium stood at under $10/lb, the lowest real price since it became a commodity and $20 below the official floor price for contract sales. The uranium industry in South Africa, the world's biggest exporter, was reported to be on the 'brink of collapse'[3] as electricity generating companies ran down stocks and cut purchases. It was not surprising that the amount spent on exploration sank from $760 million in the still relatively fat year of 1979 to between $120 and $150 million at the end of the 1980s and the size of known reserves remained static over a period of at least four years.[4] In the United States it remained the case that no new nuclear generators had been ordered since 1978. The situation was slightly better than in the first half of the 1980s, when more power stations were cancelled than entered service, but those under construction all dated from orders placed at the beginning of the 1970s. In Britain, three of four planned PWRs were cancelled at the end of 1989. In the rest of Europe the story was much the same, in some ways worse, as a glance within the hopefully green cover of the 1989 status report of the Forum Atomique Européen (Foratom) confirmed:[5]

Austria: operations at the country's first nuclear power plant, at Zwentendorf, cancelled following a national referendum.

Belgium: the government rejected as 'inappropriate at present' proposals for a new power station.

West Germany: 'the nuclear construction programme of the German utilities has practically come to an end for the time being.'

Finland: decisions deferred post-Chernobyl.

France: new orders dried up as the long run of 63 reactors either built or under construction came to an end and the country grappled with the problem of an over-supply of electricity.

Italy: no nuclear plants operating following a November 1987 referendum.

Netherlands: no nuclear power stations under construction or planned.

Spain: five uncompleted nuclear power stations under a moratorium.

Sweden: (where nuclear power provides 46 per cent of the electricity): parliament decides to begin phasing out nuclear power in 1995–6 (but parliament's restrictions on new hydro-electric power stations and the imposition of a ceiling on carbon dioxide emissions have led to a retreat from the goal of becoming, in 2010, the first country to terminate nuclear power generation).

Switzerland: parliament endorsed the cancellation of the planned Kaiseraugst power station near Basel.

The nuclear industry's current predicament is usually placed in the context of the near melt-down at Three Mile Island in 1979 and the fatal disaster at Chernobyl, the radioactive fall-out from which was still affecting places as far away as the uplands of Wales and Scotland four years later. The opposition of the activists to all forms of nuclear power, civil as well as military, had, in the view of many, been vindicated. Nuclear explosives might not have killed anyone since Nagasaki in 1945, but the use of nuclear fuel in a civil reactor had. Newspapers were full of stories of leukaemia clusters near nuclear establishments and of the long-term victims of Chernobyl and minor releases, people picked off at random within the pattern of statistical formulations. For most people in Britain, the Medical Research Council's study, published in February 1990, revealing the strong possibility that fathers who worked in the Sellafield reprocessing plant in Cumbria had passed leukaemia to their children in their sperm, moved speculation out of the realms of the circumstantial into that of the confirmed. It was a curious irony that the insidious dangers of civil nuclear power should achieve such pro-minence just as the superpowers were moving towards agreement on cutting their armouries and reducing the threat of nuclear war caused by suspicion or accident. If Three Mile Island and, above all, Chernobyl, had not happened, the image of civil nuclear power would (or should) have assumed a more friendly warmth as the superpowers' diminishing stocks of warheads dwindled towards a largely symbolic status. It would no longer be damaged by the two-facedness of its fuel.

Supposing we had grown to know and love nuclear power (as the French seem to), would we now be seeing it expanding rapidly from what the International Atomic Energy Agency claims is its present provision of 16 per cent of the world's electricity[6] to 25 per cent or more? It's doubtful. The expansion of the early 1970s was largely the result of the oil crisis, a strategic decision made in the face of economic blackmail by the Arab members of OPEC (Organization of Petroleum Exporting Countries) during the Yom Kippur War. Given the long lead-time between ordering a nuclear plant and its commercial operation, the high cost of construction, the equally awful cost of eventually decommissioning it, the margin for safety, the time-consuming process of obtaining approval (the enquiry into all aspects of the British PWR Sizewell B took two-and-a-quarter years and cost £20 million) the commitment has to be strong and 'strategic' enough – as in France – to counter the lack of resolve as the bills mount. By the second half of the 1980s oil was cheap once again, thanks to the spur the crises had given to non-OPEC developments like the Alaska North Slope and North Sea oil fields. Strategic reasons declined. The humiliating 'commercial' figures the British industry – the pioneer of civil nuclear energy – was obliged to produce in advance of privatization in 1989 practically trebled the previously estimated cost per kilowatt hour. Electricity generated by the one remaining PWR was expected to cost 8p (possibly 10p, according to one report) a kilowatt hour against 3p for power from a modern coal-fired power station and about 2.5p from one of the advanced gas turbine power stations. But these estimates were, of course, based on the short-term, very non-strategic demands of investors who wanted to be assured of a good return on their money if they bought shares in the industry. It was impossible for the nuclear power stations to match those demands.

Only the French seemed able to meet challenges of that kind, and how they did it was a mystery to most outside experts. Greater commitment, no doubt; better technology and a higher level of efficiency in construction, perhaps; and quite certainly the use of a combination of secrecy and 'creative' accountancy to maintain the image of technological triumph. In fact, Electricité de France, had a deficit of £400 million in 1989 thanks to a mixture of reactor problems and drought. The latter reduced hydro-electric power and caused the shutting down of several nuclear power stations when river levels dropped and the water in any case became too warm for effective cooling. The industry's image was further dented by the leaking of a secret report which criticized overcapacity and questioned whether its export prices would be economic if all costs were included. An added embarrassment was a report (also leaked) by the chief inspector for nuclear safety who warned against complacency and came up with an estimate that there was a 'several per cent' chance of a serious accident in the next twenty years. An accident in France would not have to be as serious as Chernobyl to imperil life and health over a wide area in France and its heavily populated neighbours.

In Britain, commercial *realpolitik* had won the day. A government committed to the disciplines of the market had had a salutary lesson: the market did not care a damn for the long term, or strategic thinking into the middle of the next century, or, quite possibly, sustainable development. The latter, in fact, was somewhere in all this a casualty. Plutonium and enriched weapons-grade uranium aside (and admittedly it is a big aside), nuclear power in its advanced manifestations is as close to the pure spirit of sustainable development as anyone is likely to get, short of covering the landscape with batteries of windmills. The fact that it is costly merely underlines the fact that sustainable development *is* costly. But in the space of less than eighteen months the government had curtailed the nuclear programme and sent the FBR into a dormancy hard to distinguish from a death coma.

Anyone who wanders into the fissured terrain of energy economics will before long encounter Keepin and Kats, from the Rocky Mountain Institute in Colorado. They are names which, mentioned favourably, are guaranteed to cause narrowed eyes and a suspicion of froth at the mouth corners in senior members of the nuclear establishment. Keepin and Kats were responsible for a calculation which has been taken up and treated with biblical reverence by the anti-nuclear lobby: for nuclear power to displace coal from the energy mix in a high energy scenario 8000 large reactors would have to be brought on line worldwide at the rate of one every one-and-a-half days. Even then, carbon dioxide emissions could still increase 65 per cent by 2030. The scenario for medium energy growth was little better: a reactor every three days. Not only was nuclear power slower and considerably more expensive than energy efficiency as a means of cutting carbon dioxide emissions, but its overall potential was much smaller.[7] 'A ridiculous cockshy to put up and then knock down,' was Collier's reaction to the Keepin and Kats paper. No one had suggested nuclear power could make the dominant contribution. It had to be balanced with other sources of energy, perhaps gas.

An entanglement of questions

There is never any one answer to any major problem and Keepin and Kats' calculations are to that extent a cockshy. It would, after all, be a lunatic commander-in-chief who relied solely on nuclear missiles for his country's defence and dispensed with infantry, tanks, aircraft and submarines. Energy efficiency may be the quickest way to abate emissions of carbon dioxide but it is hard to imagine it being applied effectively in Third World countries, whose output of the gas is likely to overtake the developed world's in the second or third decade of the next century. It may not be quite as easy as some assume in the developed world, either. Natural gas as a replacement for coal is important, too, but coal is cheap and plentiful and likely to remain the principal fuel of China, potentially the world's biggest producer of carbon dioxide (and other pollutants, too). John Wakeham, the British Energy Secretary, had put a new twist on the need for nuclear power at the September 1989 World Energy Conference in Montreal by suggesting that the developed countries might have to turn increasingly to nuclear power to 'make room' for more use of fossil fuels by developing countries.

Nuclear power is undoubtedly a big 'saver' of carbon dioxide. The South of Scotland Electricity Board generated 30 per cent of its electricity from nuclear reactors in 1980–1 and 64 per cent in 1990–1. It estimates that in that time it cut its carbon dioxide emissions by more than 20 per cent.[8] The only other countries in Western Europe which were able to achieve reductions of the same magnitude were those where the nuclear share of electricity were 50 per cent or more. A comparison of Britain and West Germany on the one hand, and France on the other, makes the point clearly. Coal is the main fuel used for electricity generation in Britain and Germany. In 1986 Britain's power stations were estimated to have produced 166.3 million tonnes of carbon dioxide, Germany 187.2 million tonnes. France, which generated 70 per cent of its power from nuclear reactors and 18 per cent from hydro-electric and geothermal sources, produced 30.2 million tonnes of carbon dioxide. Since electricity generation is the world's largest source of man-made carbon dioxide, it is quite clear that nuclear power can and does make a very significant contribution to its reduction.

In the end one plunges out of a thicket of statistics into an entanglement of questions. Which is rated as most dangerous, carbon dioxide or nuclear power? Can energy efficiency and a greater dependence on natural gas cut carbon emissions sufficiently on their own? Will low-cost uranium reserves be so depleted early in the next century that prices will begin to rise sharply? What emerges from an attempt to answer the last question is that the known low-cost reserves will have been significantly run down by 2025, but total reserves are large, sufficient to last for more than 400 years at present rates of consumption, according to one estimate.[9] (Other estimates put the span at 75–175 years.) Because of the exhaustion of the low-cost reserves, uranium prices seem bound to rise in the course of the first half of the next century (assuming the uranium mining industry is to remain in business), but by how much is anyone's guess. Until there is a substantial increase – indicating scarcity and demand – the fast reactor seems set to remain in one of the more exclusive suburbs of El Dorado.

The British study for the Intergovernmental Panel on Climate Change indicates that regardless of how fuel prices, whether uranium or fossil, change, nuclear power has a declining role in British energy production up to at least 2020. Gas could be the winner if prices are low, ousting coal. However, if one turns to the report's tables, based on a growth in energy demand of between 1 and 2 per cent a year, it makes little difference whether gas or coal is the main fuel used to produce electricity. The difference in the context of carbon dioxide emissions from all sources in the economy is estimated at between 234 million tonnes if gas is the main fuel and 250 if it is coal. As for energy saving, the study points out that the oil shocks of 1973 and 1979 cut consumption by only 12 per cent. The 'realistic scope' for saving in 2005 and 2020 is put at '20 per cent of the energy bill of the day for investments with short payback periods'. But that, the report emphasizes, is not a forecast of what will happen. The net result is that, however one mixes the energy cake, carbon dioxide emissions go on increasing. Putting the government's abandonment of the PWRs in the context of the study, the conclusion must be that it decided that, under market force rules, carbon dioxide had to be allowed to win the day over nuclear power.

Nuclear power may be in eclipse, but it is hard to believe that it – and the fast reactor – will not return to favour as carbon dioxide increases and the greenhouse effect's regime becomes more severe. Market forces will not rule in those circumstances. The health and safety hazards associated with nuclear energy will have to be overcome. Officials concerned with environmental policy predict that a day of reckoning will come when the issue has to be faced again. A combination of the ideology of the market-place, public hostility to nuclear power and fears of high energy prices adding to the problems of an uncompetitive manufacturing industry may have resulted in a setback for nuclear power in Britain, but fossil fuels would no longer seem such overwhelmingly good value, according to Chris Patten, if their wider environmental costs were allocated and more fully reflected in prices.[10] These would have to be raised to pay for the damage they caused in acid rain and global warming. As that happened, nuclear power would become more economically attractive despite the environmental costs of decommissioning redundant reactors and disposing of radioactive wastes.

The trouble with nuclear power is that its 'costs' always have the potential to go well beyond the point where they can be factored into an environmental bill. Enriched uranium or plutonium in the wrong hands could be infinitely more dangerous than an oil-rig fire or a coal mine disaster. Acid rain seems a minor hazard compared with the fall-out from a reactor fire. The danger of contamination from radioactive wastes persisting into ages which exist only as figures beyond the range of human imaginations belongs to an order different from that of oil spills and coal slurries. But if you accept that carbon dioxide is a much more insidious threat and nuclear power is the one major supplier of energy that does not produce it, the issue begins to turn not so much on the dangerous unacceptibility of reactors but on how you regulate and make them safe and ensure that their fuel is not diverted for military purposes.

Chernobyl had one good result. It led to the creation in Moscow in May 1989 of the World Association of Nuclear

Operators. The signing of its charter by representatives of 144 electric utilities with working nuclear power stations was a recognition that the time had come to take the secrecy out of nuclear power and prove that there was a worldwide interest in co-operation and information exchange. Only by ensuring that safety and reliability were maintained could public confidence be restored. WANO has regional offices in Atlanta, Moscow, Paris and Tokyo and a co-ordinating centre in London. Its chairman is Lord Marshall, who resigned as chairman of the Central Electricity Generating Board at the end of 1989 in protest against the government's decision to abandon the British PWR programme.

The IAEA and safeguards

Before a combination of Chernobyl, glasnost and the easing of East–West military tensions lifted the blanket of secrecy covering civil reactors in the Soviet Union WANO would not have been possible. It complements in a modest way the safety programmes and the 'safeguards' monitoring carried out by the International Atomic Energy Agency from its headquarters in Vienna. The IAEA's role in policing the 1970 Non-Proliferation of Nuclear Weapons treaty through its safeguards provisions is unique among UN agencies. Its international inspectorate patrols the borderline between the civil use of nuclear establishments and fuels and the threat of their diversion to make nuclear weapons. With on-site inspections, seals, analysis of materials, electronic surveillance and satellite links they cover some 95 per cent of the world's nuclear installations outside the five nuclear weapon states – Britain, France, the Soviet Union and the United States, which all allow safeguards monitoring in a few of their civil installations, and China, which has signed a safeguards agreement but has not yet had a single inspection. The non-proliferation treaty has not prevented countries like Israel, India and South Africa from secretly developing nuclear weapons, but at least they have not gone public with them and no country has officially added itself to the nuclear weapons states. Iraq, a signatory to the non-proliferation treaty, appears to have managed to divert nuclear materials to bomb-making despite regular safeguard inspections, but the treaty has been a success on the whole – as indeed has been the complex structure of hot-line, surveillance and negotiation which has ensured that the East–West 'balance of terror' never tipped over into nuclear war:

> It is clear that the present system, perhaps indeed any system that could be devised, is not foolproof, says the IAEA.[11] It is equally clear, however, that it is a remarkable phenomenon and that the chances of its leading to the detection of a diversion that has already taken place, or is in the course of taking place, are reasonably good and are improving as the Agency's technology in this area develops . . . . A top-level decision to embark on a nuclear weapons programme . . . might well involve falsification of records and a good deal of covering up. It would then be necessary for the government concerned to face the consequences of violating its Treaty obligations in such a flagrant manner – consequences which could be serious indeed.

Probably no system can ever be foolproof, particularly when developing technologies continually present new problems for the monitors. Even with the ending of East–West confrontations and the prospect of nuclear armouries being run down rapidly, it is hard to imagine a world in which civil reactors are widely regarded as safe and the problems of acceptable disposal of nuclear wastes are overcome. But internationally agreed carbon taxes, permits and rising global temperatures may push the world along the road towards accepting the unacceptable. Energy efficiency, natural gas and the limited contribution to electricity supplies from renewable resources will not be enough to contain increases in carbon dioxide and provide for the needs of a fast-expanding world population.[12]

Even if nuclear power does not become the 'sustainable' energy source of the future, the IAEA and its administration of the Non-Proliferation Treaty's safeguards regime remains of great relevance to the monitoring and application of atmospheric agreements. There will be a need at some stage for an international agency large, influential and well staffed enough to oversee and monitor measures to reduce greenhouse gases and perhaps administer aid funds. The IAEA, in its offices on the far side of the Danube beyond the Prater's great Ferris wheel, provides a prototype.

Notes

1. For an account of the early years of the FBR saga see Duncan Burns, *Nuclear Power and the Energy Crisis: Politics and the Atomic Industry* (Macmillan for the Trade Policy Research Centre, London, 1978).
2. UKAEA memorandum, *The Fast Breeder Reactor* (Energy Committee, session 1988–9, HMSO, HC-613).
3. NUEXCO Monthly Report, June 1989.
4. *Uranium: Resources, Production and Demand* (Joint report by the OECD Nuclear Energy Agency and the IAEA, 1988).
5. *Nuclear Power in Western Europe*, Status report 1989 (Foratom [Forum Atomique Européen], London).
6. *Facts About the IAEA* (IAEA, Vienna, 1988), pp. 5–6.
7. Bill Keepin and Gregory Kats, 'Greenhouse warming: comparative analysis of nuclear and efficiency abatement strategies', in *Energy Policy*, vol. 16, no. 6, December 1988.
8. Memorandum submitted to the HoC Energy Committee, Vol. II, 6th report session 1988–99 (HMSO, HC-192-II).
9. See memorandum to Energy Committe by Andrew Holmes, p. 61, HC-613, ibid.
10. *Independent*, 6 December 1989.
11. *International Safeguards and the Non-Proliferation of Nuclear Weapons* (IAEA, Vienna, 1985).
12. Even with nuclear power it will not be enough, but it might be better than not having it. To buttress its claims for nuclear power's stake in the future, IAEA points to the fact that countries like India and China plan to rely increasingly on coal to provide for their energy needs; world use of coal as expected to increase by 36 per cent between 1986 and 2000: *International Atomic Energy Agency's Contribution to Sustainable Development: Nuclear Energy and the Environment* (IAEA, Vienna, 1989).

CHAPTER FOURTEEN

The pioneering Netherlands: cows, cars and the Western world's age of mobility

The bald facts about the Netherlands suggest an environmental disaster, the country a combination of sink and sewer for the air- and river-borne poisons and wastes of Europe, including its own. The densities per square kilometre of its human and livestock populations are greater than anywhere else in the continent. Its 12 million pigs and 5 million cows (considerably more, when added together, than the human population) burden the land, and the water supply, with some 100 million tonnes of dung every year. It has more cars per square kilometre than anywhere else in the world. More than 80 per cent of the country would revert to lagoons and salt marsh if the sea defeated the country's vigilant engineers. The sea is remorseless and patient. Nature is on its side. Its level rises slowly and the land behind the dykes sinks, but even more slowly. No wonder that Queen Beatrix's 1988 Christmas message to her subjects was notable for an unfestive concentration on doom.

> The earth is slowly dying. We human beings ourselves have become a threat to our planet. Those who no longer wish to disregard the insidious pollution and degradation of the environment are driven to despair.

Strong words which no one would contradict with confidence, even if despair was regarded as an emotion which should never be admitted. But the visitor travelling through the

Dutch landscape is struck by the contrast between the royal gloom and the pleasant world all around. A heron makes a measured flight across the setting sun towards a heronry, no doubt protected by the nation's countless defenders of the natural world. The fields are a fresh dark green, the occasional unyarded cattle as majestic as any painted by Cuyp long before the greenhouse era. The cities with their canals and punctual trams are among the most pleasant and orderly in the world. The Dutch have fought hard for their environment and more than any other nation they are aware of how unremitting the battle is. Weaken, turn your back for a moment and it could be lost for good. Which helps to explain why the Netherlands' National Environmental Policy Plan[1] is a pioneering work, a model which other nations study. It is entitled *To Choose or to Lose* and it proclaims a strategy 'developed against the background of the desire to solve or gain control of environmental problems within the duration of one generation'. The plan is not a blueprint for the future, it stresses, merely the starting signal for a process in which dialogue and choices made by individuals, private industry and government will shape the environmental programmes: 'What we want to do, in the realization that we do not understand all the relationships, is to indicate the conditions under which an environmental quality can be attained that will provide future generations with as many options as possible.'

Reading the plan it is hard to see it as the centrepiece of what was billed as the world's first green election, in September 1989. Controlling ammonia emissions and ending tax-breaks for commuters who drive their cars to work are not usually the kind of issues which sends voters raging into the polling booths. Yet, it was an argument over how to pay for cleaning up pollution which in May 1989 collapsed the Christian Democrat–Liberal coalition. Liberal pique over what they felt was inadequate consultation and personal rivalry between the leaders of the parties were seen by some as the most potent causes of the rift, but the official reason which the Liberals, the junior partners in the coalition, gave was their objection to the ending of tax deductions for car commuters and the raising of taxes on petrol and diesel fuel. They supported the plan, but insisted that the money for implementing it must come from economic growth and cuts in spending on welfare and education. The voters decided otherwise and Ruud Lubbers was returned in the September 1989 election with more seats for his Christian Democrats and new coalition partners, the Labour Party. By that time the state of the economy had eclipsed the environment as the main issue. 'Ecology remains subordinate to the evolution of the economy,' commented Aad van den Biggelaar, of the environmental group *Nature and Environment*.

The plan estimates the costs of implementation in some detail – about 3 per cent of national income up to 1994, settling down to about 2 per cent by 2010. By that year, energy consumption should be down 30 per cent and emissions of pollutants by 70 per cent. Total investment would be about £45 billion. The biggest shares of the cost would go to cleaning up and limiting acidification (to which Dutch soils are among the most vulnerable in Europe) and disposing of wastes. Industry and agriculture would pay about half the bill, the government the other half. The farmers see a danger that their products will become uncompetitive; the road freight carriers have similar fears about what will happen if 1992 and the advent of the European single market exposes them to competition from foreign firms unfettered by the restrictions and burdens imposed on the Dutch. The food and drink industry and trade generally also see problems.

Fears expressed by these sectors of the economy led to the plan's drafters and the Central Planning Bureau carrying out two impact studies: one based on unilateral action, the other assessing what would happen if other EC countries adopted similar measures. The figures are relative to a middle-of-the-road estimate of the course of the economy during the next twenty years. Without similar action abroad, there is a modest decline in employment and production. Agriculture would be the biggest loser, with the volume of production cut by 10.7 per cent and employment reduced because of a drop in the number of livestock. The energy sector is the next biggest loser, hit by the decreased demand brought about by improvements in energy efficiency. Transport does well because of investment in public transport as the plan's measures to persuade people to leave their cars at home take effect. Other consequences would be worsening of the already large budget deficit and a fall in tax revenues. Most people would be worse off financially but not by a great deal. If, however, other countries take similar measures to cut pollution, the situation obviously improves. Tax revenues increase and the impact on personal spending power is less. Generally speaking, the impact seems small either way. The plan claims that the reduction in expected economic growth would be between 0.25 and 0.50 per cent a year if no other countries followed the Dutch route. That would mean the growth rate in employment would be down by between 1000 and 5000 jobs a year. On the other hand, if other countries adopted the same measures, the effect on the economy would be mildly beneficial. But trying to measure what will happen over a span of two decades is a hazardous task and the Central Planning Bureau admits to uncertainty. Dutch policies applied generally might have a damaging influence on the world economy, it says, 'How great such influence might be and to what extent it might be worse that that resulting from the worldwide deterioration in the quality of the environment is not known.' And that, in the age of sustainable development, is a fairly big caveat.

Cross-currents in the age of mobility

A bold plan of this sort directs its followers into unknown country where the statistical projections may be profoundly untrustworthy. An economic recession could throw it right off course. The loss of competitive edge might open up demands for protective measures against countries which have imposed no limitations on their own industries. But there was no sign of backing down, rather the reverse. By early 1990 an 'NEPP-plus' containing additional measures was being prepared by the Cabinet for submission to the Second Chamber. Paul de Jongh, the plan's project leader, tends to dismiss industry's worries about uncompetitiveness: 'These are partly ritual dances,' he said.[2] 'It is the first round. Private industry recognizes the virtue of the Plan because they know where they stand. I think the big businesses are prepared to adopt the policies.' He had found that even in Britain, where officials tended to be dismissive of the idea of a plan, people from private industry were receptive when he explained the Netherlands' plan at a meeting: 'They said, "Yes, we should like such a plan because it reduces our uncertainty. We like the idea of stability of policy and time to work on the reduction of pollution at our own pace".'

There are not as many cars as pigs in the Netherlands, but the country nevertheless has the greatest density of cars anywhere in the world – 128/km².[3] Britain, by contrast has 59, Japan 47 and France a mere 28. The Dutch may hate the congestion, the fumes and the wasted time brought by cars, but ownership is nevertheless as essential to their perception of what constitutes the quality of life as a bathroom and central heating. The ownership and use of cars in the Netherlands has doubled in the past fifteen years. Only 11 per cent of commuters used public transport in 1985 and the trend is downwards. Bicycles are popular, but they have their limitations on a wet and windy day. As matters stand, the number of vehicles is expected to grow from five million in 1985 to almost 8 million in 2010. The number of kilometres they cover collectively will, it is predicted, increase by 65 per cent. Goods traffic on the roads will show a similar upward trend. Roads are already crammed with traffic despite an intensive road-building programme, so presumably more of the precious Dutch countryside will be concreted over. Although passenger cars are expected to be 80–90 per cent 'cleaner' (and lorries 75 per cent) under the plan, there obviously comes a point where, unless there is an endless expansion of the road system, so many vehicles will defeat their purpose, mobility, and still be, however 'clean', a pollution problem. De Jongh admitted the dilemmas.

There is a consensus among the public that cars should not be used to go to work. But they still want them for pleasure and social purposes. There has been such an enormous increase in mobility in work and recreation. The trains are full and the roads are full. There are so many interconnections between industries and businesses and government departments. For instance, there were 400 people who were consulted in the drafting of the Plan. How did one consult them? One went to see them. [By taxi, usually, in de Jongh's case.]

The Dutch are innovators. Given their historic contest with the sea, perhaps they have to be. But what strikes one about their situation and the way they are tackling it is how vary narrowly they are ahead. Every industrialized country faces similar problems and is either considering similar solutions or has already implemented some of them. In a way national plans in Western Europe are an anomaly. They can succeed only if they fit in with similar international plans; in the Netherlands' case, with the standards agreed by the European Community to which it belongs, and that applies particularly to vehicles. Wherever one goes, whatever statistical projection is used in an attempt to bring the future into focus, one is aware of a rising sea of people and their vehicles which defeats frontiers. It is the mark of the age, the constant fretful cross-currents and surge of tidal movements between offices and homes and shops and conferences and holidays. Little more than a century ago most people, even in industrialized countries, rarely travelled more than a few miles from their birthplace. Only wars and emigration shifted them. Vagrants were returned under the Poor Laws to where they were deemed to belong. Today's vagrants, squatting under railway arches and in shop doorways, are not regarded as having strayed from anywhere. They are the only fixed residents in the shifting scene. Once academics and scientists kept in touch by letter and 'corresponding members' of learned societies were for the most part never seen. Today they spend much of their time flying to conferences, of which there are an unending supply, particularly those associated with the environment. In business and government there are countless committees and consultations and yet more conferences. Even drawing up proposals for recycling waste materials in Britain requires ten sub-committees. Faxes and telephones add to the flow of communication but do little to reduce the flow of people. Transport is cheap. Commuters regularly spend three hours travelling to and from work. Those who live close enough to walk to work make a point of mentioning it, rather as if they were keeping alive a forgotten folkway.

'We want to say "no" to all development which causes more mobility,' said de Jongh. 'It will mean very strict planning – no more supermarkets and hypermarkets, no new industrial development sites.' One idea which is to be given a test run in 1993 is road pricing. Cars will be fitted with equipment which will register them when they pass through electronic gates. It will be possible to bill the owners for their mileage. De Jongh sounded doubtful about the prospects for the idea's success. 'To be honest, I think it's a lot of fun for the engineers.'

Every planning projection of vehicle numbers and mileage has, in Britain at least and no doubt in many other countries, too, been proved wrong. The figures have soared above the modest estimates of the planners. People want cars above all else. They will wince at the idea of paying for their children's school books or having their gallstones removed, but think nothing of spending half a year's income on a car. Companies throw them in as enticing make-weights in their employees' pay packets. People may not like what cars do to their environment, but they are reluctant to give them up. If your household has two cars and you get rid of one of them you may be helping the environment, but are certainly worse off. Only the elderly make that sort of sacrifice as incapacity or reduced income takes its toll. And as the individual thinks, so, in more general terms, does the larger community: the decline of Britain's automobile industry to the point where imports exceeded exports became the measure of the country's fall from a place in the industrial front rank. Yet even in its present state it accounted in 1987, directly or indirectly, for 600,000 jobs and 2 per cent of gross domestic product.

An almost unstoppable momentum

Transport generally, and cars in particular, is such a huge business that it is hard to see its growth being checked without serious economic consequences. Since the Second World War the motor industry has gone on and on growing, undeterred even by two oil shocks which sent the price of petrol soaring. In 1950 there were an estimated 50 million cars in the world; in 1986, 386 million.[4] Even if there were a willingness to reduce production, it would be hard, in the European Community, to resist the sheer momentum created by opening the frontiers of the new internal market after 1992. It is estimated that its completion alone will increase lorry traffic across frontiers by between 30 and 50 per cent. To growth of that nature can be added the prospect of a fall in car prices, making them even more affordable than they are already. The Task Force on 'the environmental dimension' of the internal market described itself as 'much concerned' with the transport sector, which it thought would have a greater impact on the environment than any other sector.[5]

The growth in road transport presents government with dilemmas and almost irresistible temptations. They could be seen very clearly in Britain in 1989. In the first place, everyone travels by car and therefore everyone has first-hand knowledge of how awful the roads are. The great majority would like to see better and wider roads which would be less congested and safer. An *Observer*/Harris poll in 1989 revealed the strength of the British voters attachment to their cars: 'Voters overwhelmingly backed the continued, untrammelled use of private cars, with 62 per cent disagreeing with further restrictions on their use and a similar number (63 per cent) in favour of a heavy programme of road building.'[6] A government which allowed itself to be perceived as anti-car would clearly be running a risk. The moral to be drawn from polls of that sort is that spending money on roads is going to win votes. The ministers concerned must have congratulated themselves that politically at least they had done the right thing by ignoring the environmental bodies and promising eight-lane motorways and an expenditure of £12.4 billion to ensure that 1992 does not leave Britain stuck in a permanent traffic jam. The government's white paper on the subject was entitled *Roads to Prosperity* and the then Transport Secretary, Paul Channon, proclaimed, 'I will not allow Britain's first-rate businesses to be disadvantaged by a second-rate road network. Far from being the enemy of the environment, the expanded road programme will improve the quality of life for thousands of people.' *Roads to Ruin*, responded the Green Alliance, which estimates that parking alone for all the new cars expected to be on the road in 2025 would require an area larger than Berkshire. The Confederation of British Industry stepped forward with a plan that practically doubled the sum to be spent by the government: £21 billion to rewrite the nation's economic geography and save it from catastrophe. Britain's transport network pointed towards the colonial past, the CBI declared: it had to be redirected to face the European future. Anyone who has driven down the Dover Road to catch a ferry might agree.

With Japanese car manufacturers investing heavily in Britain to ensure they are inside the walls if the internal market turns into Fortress Europe, there is a possibility that Britain will become a net car exporter once again, a development which, more than any other, would help put the manufacturing balance of payments back in the black. No government is likely to say at this juncture that it regards cars as an environmental menace and intends to reduce their numbers by so many per cent a year. It is the governments of countries which do not manufacture them – The Netherlands and Denmark, for instance – which press hardest for action. The Dutch went ahead on their own in 1989 by approving tax incentives favouring cars fitted with catalytic converters (emission filters costing between £300 and £600).[7] The European Commission's protest that they were distorting competition by acting unilaterally was successfully beaten in the European Court.

Moves of that sort, though, are marginal to the general trend towards strict emission controls which will make catalytic converters compulsory in the EC. The Americans have had rigorous standards since 1983, and Australia, Japan, Norway, Sweden and Switzerland have applied similar measures. The EC reached agreement on large cars in 1987 and, after a hard-fought battle, it was agreed in June 1989 that emission standards would be applied to all new small cars from 1992. Agreement on medium-sized cars was expected in 1990. The British front in this battle has not been one of the easiest. Motor manufacturers have been sceptical about the efficiency of catalysts. They need high temperatures to be effective; stop–start driving of the sort prevailing in European cities does not suit them; the planned three-way catalysts are 'poisoned' if leaded petrol is used; they add 10 per cent to fuel consumption. It is an attitude which may go part of the way to explaining why the House of Commons Environment Committee was 'appalled' by the standard of evidence offered by the car manufacturing companies at its hearings on air pollution.[8] They showed 'an almost total lack of awareness' of the roles of nitrogen oxides and hydrocarbons in producing ground-level ozone. The Society of Motor Manufacturers and Traders made no mention of ozone at all and two of the manufacturers largely ignored it.

Vehicles are the main source of man-made nitrogen oxides, one of which, nitrous oxide, is a greenhouse gas. They are an important ingredient of acid rain and combine with hydrocarbons to form smog. Vehicle exhausts are also probably the biggest producers of man-made carbon monoxide, which destroys hydroxyl, an atmospheric 'cleansing agent' which gets rid of methane, another greenhouse gas. Catalysts are expected to cut nitrogen oxide emissions from cars to about 10 per cent by 1989 levels by the year 2006. Carbon monoxide and hydrocarbons will be less dramatically reduced, possibly by considerably less than half, while carbon dioxide emissions will increase because catalysts lead to heavier fuel consumption (only lean-burn engines which use less fuel would cut CO_2 emissions, but progress on them is slow). For all vehicles, including heavy goods lorries, the types of emissions covered by catalysts are expected to fall by 2006 to a minimum of about 70 per cent of 1989's output. To what extent is that really good news? A report commissioned by the World Wide Fund for Nature from Earth Resources Research[9] points to a lack of progress in cleaning up nitrogen oxide emissions from heavy lorries as the weak spot in the programme. As the increasing number of cars will mean that, collectively, their emissions will remain substantial, it seems unlikely that a dramatic reduction in nitrogen oxide emissions can be expected. 'On the contrary, with high levels of transport growth the emissions of nitrogen oxides from road vehicles may be higher in 2020 than at present.' As for carbon dioxide emissions, they could more than double by 2020. Even assuming a 1 per cent a year improvement in fuel efficiency from the early 1990s on and combining it with the government's lowest traffic forecast, carbon dioxide emissions could increase by 20 per cent by 2020. If energy saving is applied rigorously in homes and industry and the power stations clean up their act by switching from coal to other fuels or otherwise cutting their carbon emissions, transport seems set to take over as the biggest emitter of carbon dioxide in Western industrialized countries.

The environmental issue of the decade

Logic would seem to point to car numbers never reaching the astronomic figures produced by the analysts. The predicted 140 per cent increase in traffic in Britain by the end of the first quarter of the next century was an 'unacceptable option', said Chris Patten.[10] 'I think it's a moral issue as well as an issue of economics. It's about the future of our planet and that's something worth fighting for . . . the real challenge for politicians is to convince people that being keen on the environment is not a cost-free option.' Fitting catalysts is expected to cost British motorists £800 million a year.[11] During the 1980s, the British tended to buy bigger, high fuel-consumption cars, so increased taxes on petrol (favoured by Patten) would be an obvious way of driving them towards smaller, more economical cars with fewer noxious emissions. The Transport (formerly Energy) Secretary Cecil Parkinson took the view that cars were not going to go away and discouraging ownership would be 'very, very unpopular'. Doubling rail traffic would cut carbon dioxide emissions by only about 3 per cent. The same cut could be achieved with a 100 cc reduction in the size of engines.

We have looked at the attachment of motorists for their cars, the increasing use of heavy lorries, and the reasons why governments are reluctant to impose taxes which push up inflation and weaken an important manufacturing industry. But, despite the attractions of mobility, attitudes can be changed, or at least modified. The EC's environmental taskforce '1992' report contains the result of a poll comparing national attitudes within the Community on the importance of environmental protection. The poll is not analysed, but it is striking that countries where pollution from cars has become a major problem have the highest percentages of people who believe that the state of the environment requires urgent action: 85 per cent in Italy (at the top of the list), where Milan, one of the world's smoggiest cities, now has car-free days and the capital Rome also suffers severely from vehicle congestion and pollution; 84 per cent in Greece, where restrictions on cars in Athens are imposed to save the citizens' lungs and their ancient monuments; 80 per cent in the most vehicle emission-conscious country in Europe, West Germany, whose dead trees have become the symbol of pollution. In the United States, the strict new Clean Air Bill passed by the senate in April 1990 would cut carbon monoxide emissions by 70 per cent and a number of other gases by 22 per cent.

Apart from catalysts, more efficient engines and reduction in mileage, the only other way of tackling emissions is to change the fuel. Southern California is moving towards replacing petrol with methanol, and – an important first – limiting the number of cars a family may own; even insisting on radial tyres because they produce less dust. Hydrogen is another fuel receiving international attention. Manufacturing it requires a prohibitive amount of electricity, unless the electricity comes from cheap renewable sources, like photovoltaic cells in the Sahara, where the sunlight is guaranteed even if the water to make the hydrogen is not. Fuels made from biomass, like ethanol, from sugar cane, are used in Brazil and a few other countries, but are subject to shortages, sometimes because of drought.

And then, of course, there is the alternative of public transport, provided it is ground transport (aircraft are major polluters). The popularity of road transport has meant that railways have either had short shrift, as in Britain, where investment per kilometre is about one-third that in France, or have had to be heavily subsidized, as in France and Germany. How you view public transport depends very much on where you live. In contrast to the nationwide *Observer*/Harris poll of voters referred to earlier, 75 per cent of the Londoners questioned by Gallup on behalf of five London boroughs[13] preferred curbs on cars and better public transport to more city roads. Restraining the use of private cars would carry no disadvantages, in the view of 53 per cent, and only 4 per cent thought that discouraging people from using cars represented a loss of liberty.

Basically, there is no alternative to better public transport if big cities like London are to survive. Better trains, improved underground systems, even the large trams used in some continental cities will play their part in the future. High-speed trains in France already compete successfully with the airlines (and cars) over relatively short distances, as, for example, Paris–Lyons. In Britain, the InterCity trains have syphoned off people who would otherwise use cars, and British Rail expects to double passenger mileage between 1988 and 2020. What makes one pause when considering the role of public transport is the dispersed nature of modern society around the big cities. The car has made it that way and it would be almost impossible for public transport to knit it together in the way that the railways did in the nineteenth century. And given the general increase in mobility, would even a massive increase in investment in railways make much difference to road traffic? Chris Patten is among the sceptics: 'Even if it [investment] were to be successful and encourage a 40 or 50 per cent increase in the use of rail, it would make damn-all difference to the growth in road traffic – it would just take a few percentage points off the top.'[14] The Dutch plan envisages people living closer to their workplace and without that happening it is hard to see how they will be persuaded to leave their cars at home. But how do you get them to live within walking, cycling or tram-ride distance of work? Does the work move to them, or are cars made so expensive that people are obliged to live close to their work, as they were for much of this century? Democratic rights of use and owner-ship, a way of life and a key industry are involved. It is a highly political issue with international ramifications which we are only just beginning to tackle and which will increasingly burden the agendas of the European Community's new environmental agency.

Notes

1. *National Environmental Policy Plan 1990–1994: To Choose or to Lose* (Ministry of Housing, Physical Planning and Environment, The Hague, 1989).
2. Interview, 25 January 1990.
3. *A National Environmental Survey 1985–2010; Concern for Tomorrow* (National Institute of Public Health and Environmental Protection, Bilthoven, The Netherlands, 1989), p. 35.
4. *State of the World 1989*. Worldwatch Institute report (W.W. Norton, New York and London), p. 98.
5. *'1992': The Environmental Dimension*. Task Force Report on the Environment and the Internal Market (Commission of the European Communities, Brussels, 1989), p. vii.
6. *Observer*, 31 December 1989.
7. Three-way catalysts (the type usually required) limit emissions of carbon monoxide, hydrocarbons and nitrogen oxides by drawing exhaust fumes through a ceramic or metallic honeycomb coated with precious metals which set off a chemical reaction. Unfortunately, carbon monoxide becomes carbon dioxide and hydrocarbons become carbon dioxide and water, so there is a penalty which is added to by the fact that cars with catalysts use more fuel and therefore create more carbon dioxide anyway. The nearest thing to a complete answer would seem to be a combination of lean-burn engines (which produce less carbon dioxide) and catalysts. However, lean-burn engines produce high emissions of nitrogen oxides when operated at speed or under strain, such as a heavy load. In any case, it has been estimated that commercial introduction of lean-burn technology is unlikely until the mid-1990s. (*Source:* Friends of the Earth briefing sheet: *Cutting Pollution from Petrol Engines*).
8. *Air Pollution* (House of Commons Environment Committee report, 1988), para. 150, vol. 1.
9. *Atmospheric Emissions from the Use of Transport in the United Kingdom*. Vol. One: *The Estimation of Current and Future Emissions* (WWF, Godalming, 1990).
10. *Times*, 27 November 1989 (a report of a BBC television interview).
11. See *Air Quality* in the Department of the Environment's *The Environment in Trust* series.
12. *Independent*, 26 February 1990 (report of BBC television interview).
13. *Independent*, 28 November 1989.
14. *Guardian*, 5 February 1990.

CHAPTER FIFTEEN

Cleaning up Eastern Europe

Is there in Eastern Europe a hideous environmental prototype of what is about to happen in the developing world? There are so many similarities: poverty and governments more concerned (until very recently) with the statistics of production than with the producers; a lack of accountability among governments and officials; an absence of standards maintained through efficient inspectorates; a lack of that motivated middle class which in the West has made its strength felt in politics; a lack of effective protest by the victims, for the most part people who have come to accept pollution and filth as part of their everyday condition. Of course, the Third World has already had – and has – some considerable environmental disasters. There is Bhopal, for example, with 2500 dead from an escape of poisonous chemicals from a Union Carbide plant, and the appalling smogs and pollution of Mexico City. And in the West, too, there is a roll-call of man-made horrors, among them the explosion at the Flixborough chemical works on Humberside and the dioxin release at Seveso (Italy), which caused the poisoning of thousands of acres of farmland and the deaths of many more thousands of cattle. But Eastern Europe is different in the sustained and omnipresent nature of its pollution. It is as if governments, having been given the gifts of technology, set them loose without an idea of how or even a wish to restrain them. Production was a good in itself and therefore not to be held accountable for industry's wastes or the ill-health of workers and their families. Assuming any records were kept, pollution was as much a state secret as the artillery's ammunition stocks. Protest was silenced and it was only the new mood of glasnost in the second half of the 1980s which revealed the full extent of what had happened.

East Germany, where the lingering Protestant work ethic marched shoulder to shoulder with Leninist concepts of electrification as the vitalizing force that would change society, has the world's highest per capita emissions of sulphur dioxide and carbon dioxide. If ever there was a place waiting for the missionary preachings of the energy efficiency industry it must be there. Somehow the country has managed to consume more energy per capita than anywhere else other than rich (and notoriously thriftless) Canada and the United States and at the same time plunge into a state close to bankruptcy. The Social Democrat parties on both sides of the inner-German border agreed in early January 1990 that £72 billion would be needed to replace the brown coal generators which provide 70 per cent of East Germany's energy and clean up the water and sewage systems. A figure of that sort, roughly twice the value of East Germany's 1986 exports and imports combined, is bound to cause incredulity, but even if the figure were halved, it would still be a remarkable indication of what has happened to a country which in the first half of the 1980s claimed to have established itself among the top ten industrial nations. The coal-fired power stations may be unhealthy, but the nuclear ones are potentially lethal, as Klaus Töpfer, the West German environment minister, heard when he visited some of them. There have been scores of potentially dangerous incidents, some of them in a generator little more than a mile from the border with West Germany. One incident, in the reactor at Greifswald, nearly caused a meltdown in 1976, an accident that, occurring close to densely populated parts of Western and Central Europe, would have been a disaster of much greater magnitude than Chernobyl.

One can go on and on listing East Germany's morbid shortcomings. It requires little research to do so, largely because West German specialists and journalists rushed across the frontier and came back gasping out tales of air pollution four times worse than West Germany's, of mercury levels in the Elbe that were 250 times EC limits, of forests where 40 per cent of the trees are sick. East Germany was the first country in which fears over the state of the environment played a leading part in the overthrow of a system of government. West Germans who had not been to the East had no difficulty in believing them. They had seen the clouds of exhaust fumes spewed by the Trabants and Wartburgs which had brought escaping East Germans to the West. And it was well known, too – and there was an element of guilt here – that East Germany was West Germany's dustbin for toxic and other wastes and plain rubbish. Of the 6.6 million tonnes shipped from West Germany and West Berlin in 1989 some 700,000 tonnes were poisonous.

There is another way in which Eastern Europe can be seen as the prototype of things to come – the flow of refugees from East Germany and the German minorities in other countries to their rich brother. In January 1990 alone there were 63,000 from East Germany and 38,000 from other East European countries; 100,000 in 1989 from the Soviet Union's 2 million-strong ethnic German population. Most were economic refugees, but there were many who claimed to be environmental refugees leaving for the sake of their health. The domestic political stress caused in West Germany by such an influx (which also of course, further undermined the East German economy, since many were skilled and professional people) was one very good reason why Bonn was prepared to pour aid into its neighbour. Only by propping it up, and cleaning it up (and in the process uniting with it), was there any chance of stability. West Germany has a constitutional obligation to accept East Germans, so the parallel with Third World economic and environmental refugees is not exact. But under the general principle of rich nations having to pay poor ones to build their economies and stem their pollution it fits well enough.

The debit side of 'progress'

East Germany's environmental plight may have had more publicity than that of any other East European country, but its neighbours Poland and Czechoslovakia are almost equally high in the pollution league. The flame-belching fires and smoke of the great Nova Huta steelworks at Krakow in Poland have become a stock shot of television coverage of Eastern Europe's environmental crisis, a begrimed image of what was once an icon of progress. Pollution in such places is claimed to be ten times worse than in the West and the tonnage of sulphur dioxide said to fall every year on the average square kilometre sounds like the equivalent of a heavy snowfall. Ten per cent of Krakow's children suffer from chronic bronchitis. Nevertheless, Krakow is not as polluted as Katowice. There, the chance of dying in one's forties is twice that of an inhabitant of relatively salubrious Krakow.[1] To all that can be added the fact that Poland's economic crisis and its indebtedness (at $39 billion) is worse than East Germany's. Poland, too, is going to require tens of billions of dollars to rebuild its economy. Czechoslovakia is better off than the other East European nations and has committed 2 per cent of its investment to environmental projects, but there once again are to be found the same dreary environmental statistics of rivers poisoned, sewage untreated, sulphur dioxide deposited and trees dying, even if the figures are not quite so bad as elsewhere. Basically any report on Eastern Europe's problems comes down to much the same things: a heavy reliance on smokestack industries burning brown coal (also known as lignite) and pouring out uncontrolled noxious fumes, factories tipping their chemical wastes straight into rivers, and un-treated sewage. Only the dimensions of the statistics vary; in the case of the Soviet Union, fairly dramatically. The Soviet Union is said to contain more than 1000 towns where its own permissible levels of atmospheric pollution are exceeded five-fold or more[2] and 20 per cent of the population live in what the head of the Academy of Sciences' biology institute calls 'ecological disaster zones'. One city of a million people, Ufa, 700 miles east of Moscow, was described in *Pravda* in 1987 as having become unfit for human habitation. It is doubtful if its situation has improved since then. The heavy use of pesticides and chemicals in agriculture in Azerbaijan and the Central Asian republics is blamed for the high incidence of leukaemia, some of the worst infant mortality rates in the world, and liver disorders and cancer of the oesophagus. Nearly one-third of the citizens of Leningrad are said to suffer from diseases of the upper respiratory tract as a result of atmospheric pollution.

On the credit side, it should be said that there is now an acute awareness in the Soviet Union and the East European countries that pollution is a social and political issue and not just something that can be categorized vaguely as an environmental problem and shoved to the debit side of 'progress'. Miners have struck in the Soviet Union over pollution (in combination with other matters) and scores of green organizations have sprung up. The largest in the Soviet Union, the Social-Ecological Union, has 200 branches and is avowedly political in its approach. Poverty rather than indifference is the reason for Moscow's inaction, it seems: it was Mikhail Gorbachev who in his autumn 1988 UN speech in New York proposed that the UN should set up an emergency centre for the environment. His foreign minister, Edouard Shevardnadze, by coincidence chose the day Mrs Thatcher's greening was declared in a speech to the Royal Society (27 September 1988) to call for more co-operation in 'ecological security'. One British reaction at the time to the greening of the Kremlin shows quite starkly how much political perceptions have changed in the short space since the autumn of 1988. Michael Heseltine, the former British Defence Secretary, warned the Royal Institute of International Affairs on 23 November that the Soviet Union had identified environmental anxieties in Western Europe and the United States as offering an opportunity for mischief-making. They were attempting to use the green movement to undermine NATO's efforts to modernize short-range nuclear weapons in Europe. 'What we are seeing here is a well thought-out, carefully crafted attempt to hijack the environmental agenda for ulterior purposes,' he declared. By the winter of 1989–90 the talk was all of troop cuts and disarmament and democratization, and the rapid moves towards German reunification had made talk of modernizing short-range nuclear weapons to be fired from West Germany at targets mainly in East Germany more plainly than ever the nonsense it had always been. President Gorbachev's green fifth column could be returned to the fantasy-land it came from.

At this point it is worth taking a step back from the scene

and recalling that the Soviet Union and all the East European states are signatories to the Convention on Long-Range Transboundary Air Pollution. With the exception of Poland and Romania, they are pledged under the 1985 Protocol on Sulphur Emissions to reduce their sulphur emissions by 30 per cent by 1993. It is unlikely that when they signed they could ever have had the slightest hope of achieving such a reduction. There have been improvements here and there in the East European countries, but generally speaking they distribute sulphur dioxide to their own fields and cities and to the countries downwind of them as liberally as they ever did.

An injured party can complain against its neighbour under the convention, but it has no teeth. What can have teeth, of course, even if it is concealed by a friendly smile, is aid. By early 1990 the European Commission was co-ordinating aid to Poland and Hungary, the two East European countries regarded as having satisfactory democratic credentials. The packages were worth £600 million each. The Japanese were quick to promise similar sums. Under consideration was extension of the EC aid programmes to Czechoslovakia, East Germany, Romania and Bulgaria. The Soviet Union was too big a problem even to begin thinking about seriously at that stage; a military superpower still, even if its economy was falling apart and its politics were in uncertain flux from one-party dictatorship towards a multi-party democracy.

The proposal by the President of the Commission, Jacques Delors, in Strasbourg on 17 January 1990 that the EC should treat the six East European countries as if they were backward regions of the Community and extend structural support to them brought gasps of disbelief from every government except West Germany's. What Delors had in mind was aid worth £10 billion a year for ten years. Only two years earlier the Twelve had battled at epic length over increasing the EC budget for the years to 1992 from 0.7 per cent of each member state's gross national product to nearly 1.2 per cent, with the ceiling fixed at over £30 billion. Delors' proposals were clearly out of all proportion to that figure, and he was soon obliged to retreat with a lame explanation that he had been 'arguing from the absurd to show what it would cost to achieve such a goal'. Nevertheless, it is clear that very large sums of money will have to be spent to put Eastern Europe on its feet and ensure that it stays democratic. In the course of doing so money will have to be spent on cleaning up the Six (and, no doubt, eventually, the Soviet Union, too) to ensure that they make respectable cohabitants of the new Europe which various leaders from Gorbachev to Delors to François Mitterrand have envisaged. Two EC countries (The Netherlands and Denmark) already examine the possible environmental impact of aid to Eastern Europe before approving projects. Britain's Overseas Development Administration has its *Manual of Environmental Appraisal* which will presumably apply as much to East Europe as it does to the Third World. The EC itself has not attached environmental conditions in the cases of Poland and Hungary, but Carlo Ripa di Meana, the EC Environment Commissioner, has insisted that a sum (still to be agreed) must be set aside to bring the East's environment up to scratch and there are proposals for training environmental managers. The new Bank of European Reconstruction and Development will obviously have an important role in deciding policy when its opens its doors for business.

Aid: East Europe vs. the South?

While the EC was debating its approach to the problems of Eastern Europe the president-elect of another distressed part of the world was nearing the end of a pre-inaugural tour which took him to all the major capitals. The message Fernando Collor de Mello brought with him was that Brazil, where he would take over from President Sarney in mid-March, no longer wanted to be tagged as a Third World nation. It wanted to be seen as an honorary member of the developed world experiencing the sort of hard times you might expect in a country with the world's biggest foreign debt, $115 billion. Under Collor's guidance, it would modernize itself. As host of the 1992 UN Conference on the Environment and Development, he hoped to see the 'depoliticization' of what he called (in London, in February 1990) the unfolding 'ecological drama':

I will endeavour to combat the political, ideological and

even electoral exploitation of the issue. This is a serious issue which must be dealt with on a rational basis. Hence the need to avoid mutual recriminations, facile accusations and scapegoats. Let us not be simplistic about things that concern the survival of mankind.

Even allowing for a desire to please his audience, sentiments of that sort sound better news than warnings of confrontation. An aspirant to membership of the developed club is likely to be accommodating about rainforests, the welfare of Amazonian Indians, the ozone layer and a climate convention. Collor has spoken favourably of the idea of an international carbon tax. But assuming Collor's Brazil makes it into the ranks of the developed world, it will do so at the lowest level of eligibility, on a par with the East European countries. The country's population is over 140 million; the total population of the East European Six is rather more than 110 million. Brazil had 83 cars and 84 telephones for every 1000 people in 1985;[3] East Germany had 238 and 218 and Poland 119 and 113, respectively. The ratios may have changed, but probably not by all that much, and there are no doubt more sophisticated and precise ways of measuring relative wealth. But it is reasonable to assume that Eastern Europe is still considerably better off than Brazil, which in early 1990 was reported to be $6 billion in arrrears on its $115 billion debt and facing the prospect of 2200 per cent inflation during the year. In 1989 Brazil repaid nearly $800 million more to the World Bank than it received from it. So it was not really surprising that despite his upwardly mobile outlook, Collor attended the paupers' summit in June 1990 at which fifteen debtor nations from the underdeveloped South discussed their predicament.

That predicament has undoubtedly been accentuated by events in Eastern Europe. The attention of all the principal aid donors is concentrated there. If Eastern Europe gets massive assistance it is logical to assume that the South will get less. Poland's gain may turn out to be India's loss. It may be essentially racist, but rich Europeans are likely to put the needs of poor Europeans ahead of those of poor Asians. The West has a great political interest in seeing democratic systems based on market economies develop in Eastern Europe, since a collapse into worse poverty and anarchy would be a disaster, a tragic first chapter to the new peace between East and West. Apart from that, the prospect of investment and increased trade with countries with long traditions of industry and commerce is a stimulating one. The return on money spent or lent in Eastern Europe can be expected to be better, politically and commercially, than those from similar outflows into the gurgling sink of Latin America and other parts of the South. It was no wonder that some politicians promoted the idea of a 'Marshall Plan' for Eastern Europe. Senator Robert Dole, the Republican leader in the US Senate, went further than most by calling for aid to be diverted to East Europe at the expense of long-standing recipients such as Egypt.

As expenditures on arms decline with the ending of the cold war, there may be more money available for aid, but democratic politics being what they are, few would care to bet on it; certainly not on any percentage saved being set aside for the developing nations. Even if it were, its benefits would be quickly swallowed by the rapid increase in the South's requirements as its populations grow; and the experiences of the 1970s and 1980s in which aid and loans failed to reverse economic decline are not encouraging. Disillusionment and the heavy losses sustained by Western banks are reflected in the fact that the total net resource flows from the European Community and the eighteen member states of the OECD's Development Assistance Committee (the world's richest countries) dropped from $128 billion in 1980 to $103 billion in 1988.[4] Eastern Europe and the Soviet Union could be equally disillusioning, of course, and when the Governor of the Bank of England, Robin Leigh Pemberton, spoke to the Overseas Bankers' Club in February 1990 on lending to East Europe, he put his call for caution in the context of the losses sustained in the South. Don't get stung again, was his basic message.

'Green conditionality'

The bankers will take the risk of getting stung again, of course, because that is the way of the banking world. They will be under political pressure to lend and there will be inducements to do so. So it is fairly safe to assume that assistance from Western sources, official and private, will flow to the East in a crusade against the poverty and pollution caused by seventy years of ideological perversity. The parts of the world – the Third World – which may suffer a loss, real or relative, as a result are generally estimated to have a debt of $1300 billion. Loans and aid of all kinds to the Third World actually dropped slightly in real terms in 1988, from $96.9 billion in 1987 to $96 billion adjusted to 1987 prices and exchange rates. Among the seventeen heavily indebted nations (mostly Latin American and including Brazil) per capita income in 1988 dropped by 0.6 per cent.[5] Sub-Saharan Africa continued its decline in per capita income, too, for the same reason as the heavily indebted countries – population increases continued to outstrip modest rates of economic growth. In fact, the World Bank believes that aid to black Africa (now running at about £10 billion a year, the same figure as that which caused Community jaws to drop when suggested by Jacques Delors for Eastern Europe) needs to increase by nearly 50 per cent by the end of the century.[6] The horror stories of Eastern Europe can be more than matched from the Third World. One in three Peruvian children is stunted by malnutrition, said the Worldwatch Institute of Washington in a 1989 report, *Poverty and the Environment: Reversing the Downward Spiral*. Infant mortality doubled in Zambia during the first half of the 1980s. Life expectancy fell in nine African counries. Almost a quarter of the world's population – 1.2 billion people – were reckoned to be too poor to meet their most basic needs for food, clothing and shelter.

It is not a uniform picture. The two biggest Third World countries, China and India, were, by contrast with Brazil, doing well so far as growth rates are concerned. They are low-income countries and India has more desperately poor people than anywhere else in the world, but in 1988 China's growth rate was 11 per cent and India's 8 per cent, well ahead of their population increases. China's growth is fuelled largely by coal and so is India's, some of it in the latter case thanks to World

Bank funding. And burning coal, of course, produces, carbon dioxide as well as sulphur dioxide. The World Bank was last year reported to be in the throes of an anguished debate[7] between its environmental unit and its energy department over the best way of enlisting Third World support for curbing greenhouse gases. Can you really tell countries like India and China that they should switch from coal to some other, more expensive fuel or spend heavily on cleaning up emissions? The Bank stuck to a policy of recommending that subsidies should be ended, which would have the effect of increasing electricity prices and encouraging energy efficiency. Proposals for a worldwide carbon tax and the phasing out of coal-burning generators appear to have been ruled out.

The subtraction from 'traditional' aid flows implied by Western Europe's impassioned concern for East Europe has worried the Third World, as its diplomats make clear every time they find a receptive ear. Even before the upheavals in Europe they were warning against attempts to subtract environmental aid from development aid. Western politicians like Lynda Chalker, the British aid minister, have tried to reassure them: 'Britain will continue to respond to the needs of the developing world,' she said in mid-January 1990. 'The very substantial funds which we are making available for Eastern Europe . . . are separate from and additional to our regular aid programme.' But statements are unlikely to prevent a heightening of suspicions which are always fairly strong even at the best of times, and that could make it harder for the North to get the global climate convention it is seeking. The South does not, for a start, like the idea of 'green conditionality' being attached to aid agreements. Such conditions were a ruse by the North to make the South pay an unfair share of the cost of protecting the environment, said Bernard Chidzero, the Zimbabwe finance minister, at the September 1989 Commonwealth finance ministers' meeting in Jamaica. The final communiqué recorded the ministers' view that aid totals would have to include 'any additional burden on developing countries arising from action to protect the environment'. That general attitude of the South's is said to have been the most striking impression retained by Ripa di Meana, the EC environment commissioner, from talks during his first year in office (1989) with Third World ministers and officials.

The agreements on protecting the ozone layer are often pushed forward as pioneering achievements which will ease the way to an atmospheric convention and protocols dealing with carbon dioxide and other greenhouse gases. They are indeed considerable achievements, but what they show most clearly is how hard it will be to get the major Third World countries to agree on any protocol which could slow down or affect industrial growth. Aid and technology transfers are their price. Poverty is a more pressing problem for them than the greenhouse effect. That came across very strongly at a UNEP conference in Nairobi in 1989 on phasing out CFCs.

Buying out CFCs in the Third World is within the West's means, but buying out carbon dioxide in China, India, Brazil and the lesser parts of the Third World is certain to prove beyond anyone's means. If persuasion does not work, 'green conditionality' might be the answer in some cases, unpopular though it would be; and for the really hard cases, sanctions, provided there is sufficient of a consensus to back them and the industrial world has not sunk into a mood of profound fatalism about the impossibility of stemming the greenhouse gases. Eastern Europe and the Soviet Union are object lessons about the dangers of letting industry befoul the environment in the name of economic growth. There is still time for much of the Third World to be held back from taking the same course.

Notes

1. Television and press reports on the state of the environment in Eastern Europe have been full and frequent. I have relied mainly on the *Independent*, the *Guardian*, the *Times* and the *Economist*.
2. *Economist*, 4 November 1989.
3. These and other figures from *World Statistics in Brief* (UN, New York, 1988).
4. From *General Background Note on the OECD's Development Assistance Committee* (OECD, Paris, November 1989).

5. *Annual Report 1989* (The World Bank, Washington, DC).
6. *Sub-Saharan Africa, from Crisis to Sustainable Growth* (World Bank, 1989).
7. See *Guardian*, 7 August 1989.

CHAPTER SIXTEEN

1992 and beyond

When it comes to handling the environmental crisis there is undoubted strength but also a flaw in the Western democratic system which has triumphed so dramatically, both ideologically and materially, at the end of the twentieth century: its political and economic well-being depends on growth and the prospect of ever-increasing wealth and improved standards of living. Governments which wish to survive have to maintain growth. Since governments with a death wish are rare, growth which enriches the individual is sacred. Nothing is allowed to impede it. The health of democracy depends on it. Depressions are death. It is a way of thinking which means that in great emergencies such as war, sacrifices are usually made late and reluctantly. There is no reason to believe that the environmental crisis will be treated any differently. Raising the sums which will be required in the name of global environmental security during the next 50–60 years (which is about as far as anyone can reasonably try to foresee) will create severe political and economic stresses. What they will add up to is anyone's guess at the moment; inflation, uncertainty about the pace and impact of climate change, and the political factors which govern priorities make any attempt at assessment a futile exercise. It is enough to say they will be counted in trillions (millions of millions) of dollars. Some recent estimates of the cost nationally in the United States give an idea of the probable scale: curbing carbon dioxide emissions at 1990 levels could absorb between 1 and 2 per cent of GNP, with the sums rising into the trillions in the next century. There is not a US coastal state which is not considering measures to preserve its coastal regions and the Environmental Protection Agency has put the cost of protecting cities against a one metre sea-level rise at between $73 and $111 trillion.[1] Other vulnerable countries face heavy expenditures, cancelling out any savings made in defence budgets – around one trillion dollars a year globally – following the end of the cold war and perhaps rising to the levels of expenditures experienced in a war. The temptation for the rich nations will be to avoid or limit them by deferring expenditures on pre-emptive measures which will add up at home to taxes, higher energy costs and technology transfers to poor countries striving to compete with the donors.

There is a taste of the difficulties ahead in the ordering of priorities in the advanced industrial countries: first place goes to economic growth and second place to environmental security. Unless there is growth, it is argued, the rich will not be able to provide the money and technological resources needed to counter climate change globally. Thus, the ozone layer can be saved only if there are substitutes for CFCs which enable manufacturers to continue making refrigerators. The fashionable creed of sustainable development panders to that sort of thinking. It tells electorates that there can be continual growth and no one need worry that the cost of environmental security will hit spending power or reduce standards of living. Governments point to just how wrong the environmentalists were in the 1970s when they claimed that the 'squandering' of resources would impose limits on growth. The 1970s and 1980s have been decades of unparalleled growth. But that is an increasingly irrelevant justification for present attitudes. The argument in the 1990s is not about resource depletion but about changes in climate which will concern life and health as well as, ultimately, wealth. Every qualified projection of climate change and population growth indicates a crisis which will grow in seriousness through the first half of the twenty-first century.

Concern with growth is not a Western/developed world monopoly. Growth is a desperate necessity for China and India, which between them account for a third of the global population, and other developing nations. The idea of diverting even one per cent of GNP to curtailing carbon dioxide emissions would seem to most of them too frivolous to be seriously entertained, unless, of course, someone else was willing to pay. And which countries, one wonders, would be willing to finance the replacement of China's abundant coal as the fuel for its power stations with less polluting natural gas, or underwrite double-glazing in Chinese homes as as major contribution to energy efficiency? Perhaps they will one day, but at the moment it seems some way off.

Taxing the air

There is no doubt in the mind of Mostafa Tolba, UNEP's executive director, that 'financing solutions to the global environment crisis will be the major issue of this decade'. Tolba belongs to a political species that will become more common as the decade rolls on towards the next century. He is an environmental statesman. When he speaks he is listened to with well-earned respect, particularly in these preparatory years before the 1992 UN conference on the environment which will mark the culmination of his career. The environmental crisis will, he says, require an increase in the flow of resources to the Third World, debt relief, fair pricing of commodities, land reform and help for Third World food producers in the form of a phasing-out of subsidies to agriculture in the developed nations. A genuine global partnership must be struck between rich and poor in which all contribute and all have an equal say in the allocation of resources. 'The worst mistake that could be made is for the rich to apply conditionality to new aid and loans.'[2] Saving the world will not be cheap, but, then, security never has been cheap. The alternative to North–South co-operation based on aid will be conflict and the prospect of '21st century eco-wars'. Could Egypt, he asks, allow nations sharing the watershed of the Nile to disrupt the river's flow, the lifeblood of its economy? How would Niger and Nigeria react to the plans of upstream nations for barrages and irrigation schemes which might dry up the Niger? These are not isolated issues. Some 40 per cent of the world's population depends for drinking water, irrigation and hydro-electric power on 214 river systems shared by two or more countries. 'The threat from a nuclear war forced us in the post-war period to think in terms of war destroying the whole planet. Human economic activity forces us to think of the environment in the same terms.'

The message is a forceful one, strengthened by the fact that resource wars are not exactly new. The search for better grazing prompted many of the great nomadic outpourings from Central Asia, and as recently as the early 1960s a civil war was fought in what is now Zaire over the control of Katanga's copper mines. However, one can accept the message and still wonder how Tolba proposes raising the money to buy peace. He agrees that 'practical and innovative thinking' is required, and he clearly likes an idea inspired by the sustainable developer's manual, users' fees. This, he says, is an 'attractive option'. It would mean pricing and paying for air and other environmental resources. Fees would be collected by national governments and a proportion of them paid into an 'international fund to save the earth'.

The idea of taxing what most people regard as their birthright, fresh air, is startling, but perhaps Tolba is right: we should not take it for granted. After all, we pay for piped, chlorinated water, why not for cleaned-up air? However, acceptance of the idea would still leave open the question of how air is to be priced. If, for lack of a formula, British ministries (and no doubt ministries elsewhere) are unable to price the amenity value of landscapes threatened by development, how on earth would any government – or, more probably a conference of governments – be able to arrive at a system for pricing the air? How much more would the rich Swiss pay than the impoverished Vietnamese? Or would it be like the British poll tax, with rich and poor paying the same subject to a complicated system of rebates? If payments were partly based on per capita emissions of greenhouse gases, would methane from paddy fields be given the same price as carbon dioxide from power stations, cars and burning tropical forests? Suppose, though, that Western countries agreed to the idea of users' fees based on notional valuations of the atmosphere and other resources. Suppose, too, that they managed to sell the idea to their tax-payers, who, after all, would have to find the money. They would then be required to hand over the proceeds, without any strings attached, to a fund in which, under the Tolba plan, 'all countries would have an equal say in resource allocation.'

Listening to Tolba produces an uneasy feeling that perhaps Nairobi, where UNEP has its headquarters, is a bit isolated, too far away from the 'real world' of money and carbon dioxide and hard-nosed political deals. The case for placing UNEP in the Third World is plain enough – influence, providing a sense of shaping the decisions – but where can the pay-off be seen in Africa, a hapless continent which does nothing to strengthen UNEP's authority? If UNEP is to play the key role in shaping and managing the new environmental conventions Brazil would be a better centre (and Geneva an even better one). Realistically, though, there is no chance of extracting UNEP from Nairobi. The politics of the United Nations will ensure that it stays there.

Tolba's ideas carry echoes of the argument in the 1970s over the New International Economic Order in which the Third World lined up to demand compensation in the form of aid and technology transfers for the 'looting' of their resources by Western imperialism. The huge Third World debt incurred as the Arab oil producers' profits from the oil boom were recycled by Western banks was the outcome. This time, the Third World has a lever of sorts in its willingness, or unwillingness, to co-operate, but one of limited value since co-operation in drawing up a climate convention is ultimately as much in the interests of the poor nations as it is in those of the rich. It is certainly not strong enough to extract unconditional aid from the developed nations, habitually suspicious of what happens to their money once it reaches Third World administrations. Donors will obviously need to be sure that environmental aid is administered efficiently and spent according to agreed priorities.

An end-of-century shibboleth

The 1992 UN Conference on Environment and Development will be a test of the world's commitment to take action to

protect itself. A climate convention (with or without protocols) should be ready by then to form the agenda's
centrepiece, and possibly there will be a convention on bio-
diversity which would cover the protection of the rainforests
and the multiplicity of species which they harbour. Carbon
dioxide will be the supreme test of commitment and anti-
pollution technology, but, as has been noted before, it is so
omnipresent, its natural cycle so great and its sources and
sinks so difficult to determine accurately (one assessment in
early 1990 downgraded the amount mopped up by the oceans
by 50 per cent but came to no firm conclusion about where the
unaccounted for remainder went), so closely linked with
economic growth and its man-made sources so hard and
expensive to restrain that it is difficult to imagine a protocol
which will be globally effective. Methane, which contributes
18 per cent of the greenhouse effect, presents even greater
problems. Curbs on ownership of water buffaloes and other
ruminant emitters do not seem likely, nor would it be possible
to cover up rice paddies and melting permafrost in the tundra.
But even if by some miracle effective protocols are agreed,
the momentum of climate change means that whatever action
is taken to reduce greenhouse gases will be of more value to
generations beyond the mid-century than to those which come
before it. Mitigating the causes of the crisis has to continue, of
course, but the dominant issue may turn out to be crisis
management within the triple-P of problems caused by
poverty, pollution and population.

The philosophy or, perhaps more accurately, the frame of
thinking which will guide governments and international
institutions in the 1990s and beyond is still in its formative
stages. The Brundtland Commission may have made sustainable development the end-of-century watchword, defining
it as 'development that meets the needs of the present without
compromising the ability of future generations to meet their
own needs', but even Tolba has wondered aloud whether it
amounts to much more than a shibboleth. Richard Saunher of
the Organization of American States' Department of Regional
Development informed the April 1987 Sustainable Development Conference, in London, that 'a concern for sustainable
development certainly does not consciously guide our work.

Indeed, we make an effort not to use the phrase for a number of reasons, one of which is that we have been unable to figure out what it means'. More importantly, it was regarded as a distraction from the real issues of environment and development. Timothy O'Riordan, of the University of East Anglia, thinks sustainability might be accepted as the 'mediating term' between developers and environmentalists,[3] although he leans to the view that it will eventually languish as a 'good idea' which cannot sensibly be put into practice. The arch-priest of the British sustainable developers, Professor David Pearce of the London Environmental Economics Centre, connects it with the 'quality of life' and people's expectations. 'Sustainable development involves devising a social and economic system which ensures that these goals are sustained, i.e. that real incomes rise, that educational standards increase, that the health of the nation improves, that the general quality of life is advanced.'[4] In both the Brundtland and Pearce reports the principle is established that present development must not jeopardize the future. Pearce encapsulates it as each generation ensuring that it passes on to the next an undiminished stock of assets, including environmental as well as man-made capital. That might just be possible in countries with zero-growth populations, but in the fecund Third World the result would be the equivalent of the ancient Saxon custom of gavelkind, in which the inheritance was divided equally among all the sons. Everyone tended to get poorer as the lots grew smaller and smaller.

Without being too sophisticated about its finer points, sustainable development has an obvious, sensible meaning when expressed in terms of a traditional farmer recycling dung and compost to maintain the fertility of his land. Like the farmer, humanity should try to live within its means. If energy can be obtained from tides, waves and windmills that is better than energy obtained from fossil fuels, which are non-renewable. However, the world's preponderance of very poor people want to get richer as quickly as they can and are probably not too concerned if they do so at the expense of the environment. Not only that, but the number of people in that category is rising very rapidly. Between 1985 and 2025 the population of the developing world is expected to increase from 3.7 billion to 6.8 billion, of whom well over half will live in those hotbeds of aspiration and discontent, the cities. Sustainability in those circumstances would seem virtually impossible. One of Pearce's main points is that we need to be able to apply monetary values to environmental gains and losses. An improvement in environmental quality is also an economic improvement if it increases social satisfaction or welfare.

We must learn to recognize that environmental capital is just as much capital as man-made capital. Environmental capital includes not just the stock of oil and gas, coal and minerals. It also includes the ozone layer, the protective functions of forests and wetlands, the waste-assimilating functions of rivers and oceans, and the store of biological diversity. To ensure sustainable development, we must 'maintain the capital stock'.[5]

Dr Pearce is the special adviser on environmental economics to Chris Patten. Patten believes that people have to recognize that environmental quality is not a cost-free option and that the way in which energy and transport policies will have to be modified to take account of environmental factors will become an important subject of public debate. So let us look at an English case which exemplifies some of the practical problems confronting the theorists of sustainable development: Twyford Down (near Winchester) and the route of the M3 motorway in March 1990. Twyford Down is a much-loved, fiercely defended 'environment' close to a city which values its green spaces. It is designated as an area of outstanding natural beauty, the exquisite chalk-hill blue butterfly has made its home there and there are Celtic fields and an Iron Age fort. Here, clearly, was a marvellous opportunity to give Pearce's principles free rein and see what costings they produced.

What had to be decided was whether 2.63 miles of the extended six-lane M3 would be tunnelled through the downland chalk or just cut straight through in a deep, wide trench which would obviously make an irreparable mess of the down. Tunnelling was estimated to cost £128 million; cutting a trench £36 million. The £92 million difference between them was plain enough, but how much was the unspoiled beauty of Twyford Down worth, not just to present-day citizens of Winchester but to future generations? The authors of *Blueprint for a Green Economy* express the belief that it is possible to put a money value on the benefit of preserving the rare Californian condor, so why not on chalk-hill blue butterflies? On harebells, primroses, rabbits and just the sheer pleasure people get from walking on such familiar, ancient places, too. £10, £20, £50 million? It is sad to report that neither the Transport Department nor the Environment Department (they made a joint decision in favour of the trench on the grounds of cost) made any attempt to value Twyford Down, its inhabitants and the enjoyment gained now and in the future from them. 'At the moment there are no formulae for costing environmental matters of that sort,' a spokesman for the Department of Transport informed the author.

It would, of course, be very difficult to create a formula. A lepidopterist might be able to quote a going price for a preserved chalk-hill blue in good condition, wild rabbits have a price when shot and skinned, but the market price of Twyford Down itself would be low if sold with legal constraints which kept it as an open, public space and high if sold with planning permission for houses or industry (or even a road). If it is so difficult to value a small piece of the environment like Twyford Down, how will it ever be possible to cost the ozone layer? The answer must be that trying to give values to the environment is an irrelevance. Perhaps one can produce figures for the amount knocked off property prices by aircraft noise and excessive pollution (as *Blueprint* suggests, with the backing of figures from American cities), but even here there is room for doubt. Some of the highest property prices anywhere are in fume-ridden Manhattan and in cleaner but nevertheless congested central Tokyo. Airports attract hotels and businesses; they employ large numbers of well-paid staff who like to live near their work. House and land prices tend on the whole to go up rather than down. There is another problem when it comes to applying market prices to environmental capital: the pieces of green environment that are most treasured by most people are their gardens. Thousands of millions of dollars, marks, francs, pounds and every other currency of substance must be spent every year on plants and equipment for them. But they are man-made capital rather than environmental capital. In the end environmental issues come down to political decisions on what people want. The Green Belt around London has been preserved not because someone put a price on it but because voters in the Home Counties wanted to stop the city sprawling outwards seemingly for ever. They wanted planning and an attempt at a balance between the traditional English landscape and the demands of housing (and gardens) and roads, quite simple concepts, really.

There was another way of tackling the issue of Twyford Down. The Departments of Environment and Transport could have agreed that in principle it was better not to encourage more cars onto the roads by building better motorways. What was needed was a much improved rail link to Southampton, backed by a tax policy that would penalize motor vehicles and favour the railways. But, again, no one seems to have considered that as an option; and if they had, it would have been highly unpopular and, if implemented, its effects slow and perhaps ultimately unsatisfactory. 'We are not going to do without a great car economy,' declared Margaret Thatcher in March 1990 in the course of dismissing 'airy-fairy' green ideas about a return to village life. 'Much of our economy would collapse if we did without that.'[6]

What sustainable development really represents, one suspects, is a dilemma. Essentially it comes down to the age-old question which has tormented the rich as much as the fate of their souls: how do you have your cake and eat it? Growth at almost any cost has always been the answer. Industrial society has managed to bake bigger and richer cakes for most of the past two centuries, suffering the occasional bouts of indigestion, but getting fatter most of the time. In the West we have bigger, warmer homes, more amusement, more communications; we travel more and further for work and pleasure than every before. Compared with our ancestors plodding six days a week between plough furrow and hearth and, taking mileage as a measurement of space, we must occupy a thousand, perhaps several thousand times as much space. Sustainable development consoles us with the idea that we can go on having more provided we are more hygienic and respect nature. We have to be vaguely provident, but no real sacrifices are demanded.

Common sense may tell us something different: that there are limits to growth (or more accurately, perhaps, to real disposable income), imposed not so much by the depletion of fossil fuels and mineral which worried the Club of Rome, but by numbers of people and the cost of a worldwide defence of the environment. Technology and resource transfers to help and persuade Third World countries to combat the effects of global climate change potentially could impose enormous charges on the developed nations in addition to what they have to spend nationally and regionally. If, for example, lean-burn engines are developed by Western manufacturers as the best way of cutting carbon dioxide emissions by the world's rapidly increasing stock of vehicles, is the technology passed on free of charge to developing countries beginning a build-up to mass production from their own motor industries? Patents and the rights of private industry are not matters which can be brushed aside cheaply in order to enlist Third World support, as the difficult negotiations on the technology for manufacturing CFC substitutes have shown. If the developed world wants to save the destruction of tropical forests by fuel-wood gatherers does it help the countries concerned to provide alternative energy from power stations using shrinking supplies of relatively clean natural gas while it switches to nuclear power, despite its expense and dangers?

Building the security regime

Sustainable development will have value if it provides politicians with a non-threatening idea to hang on to while they sort out their approach to the problems created by climate change. They know what they have to *avoid*, another Law of the Sea marathon. What they have to *do* initially is agree a climate convention. If it is to be more than a symbolic marker of the moment when North and South decided in principle to work together for mutual survival, a number of decisions on how to administer it will have to be made. Some of the grander ideas have envisaged an all-embracing law of the environment, an entirely new international environmental protection agency and even a special environmental security council. Those who argue that creating such a structure would take years and waste valuable time are probably right. The UN Charter would have to be amended to provide for a special security council which would not, presumably, include the rights of veto held by the five permanent members of the existing security council. What is best in this case is undoubtedly what will be quickest and simplest. The Inter-governmental Panel on Climate Change, which links the World Meteorological Organization and UNEP and has its secretariat in UNEP's offices in the UN's second city, Geneva, could be institutionalized and strengthened under the aegis of UNEP to provide the focal point for investigation, the monitoring of the impact of climate change, and ideas. UNEP, it is worth recalling, was born out of the 1972 Stockholm conference; the 1992 anniversary conference will be the time to consider turning it into a full-blown UN agency with more funds at its disposal. Conversion into an agency would introduce a need for detailed thought about its functions and its future role. Monitoring the climate convention and the impact of climate change would require a mandate and an establishment at least equivalent to that of the Vienna-based International Atomic Energy Agency, which polices the Nuclear Non-Proliferation Treaty. The Nairobi headquarters could stay where they are, but the climate side of its activities would need to remain in Geneva, close to the WMO and in an international hub-city.

The environment is already a hazy area which overlaps into a great many administrative realms. It will become even less distinct as environmental issues become more and more merged with problems caused by rapidly expanding popula-tions, poverty and the visibly harsh strictures of climate change. If one had to make a guess, it is that within another ten years or so 'environment' will have become a somewhat *passé* term, rather as 'ecological' has, simply because of its insufficiency as a generic description; a term which links the preservation of rural landscapes in Europe to the fate of millions in Bangladesh obviously has problems of definition. Under the cover-all of climate change, the issues will tend to become more sectorized, too big to be handled monolithically.

International organizations like the World Bank, the Food and Agriculture Organization and the UN High Commissioner for Refugees are already involved in issues and situations which are either directly or indirectly 'environmental' and their workload will increase. It would be virtually impossible to bring them under the control of one super-agency. The World Bank has already put forward proposals for a global environment fund with which UNEP and the UN Development Programme would be associated.[7] Its mandate would include technology transfers and the encouragement of investment in projects which would enhance or safeguard the environment. Rather than establish a separate organization to administer the fund, it would probably be better to leave aid of that nature to the World Bank and bilateral and regional agreements, the more so since an issue like controlling carbon dioxide emissions may come to depend not so much on targets and dates agreed under a protocol to the climate convention as on the speed with which programmes to introduce cleaner forms of energy (possibly including internationally supervised nuclear energy) can be funded and implemented.

Even if no super-agency is established to deal with the environmental crisis, it will still be necessary to create a regime which can monitor and police compliance with agreements. It should be possible to take breaches of agreements before the International Court of Justice or, when they amount to a threat to peace, before the Security Council. Obtaining compliance will be as difficult as it always has been, but an international community which has used sanctions to encourage change in southern Africa should be able to tackle the much more serious issues of enforcement associated with climate change and international security. Warnings of eco-wars and the flight of hundreds of millions of refugees cannot be dismissed as fantasies: they are probabilities in situations which become more and more apparent as the century nears its end.

Man is a god in ruins, thought Emerson, and perhaps at the end of the twentieth century much the same could be said of his world, a still beautiful but ravaged paradise which, regardless of the tenets of sustainable development will not be passed on to the next century in better or even the same condition, in fact, almost certainly in worse condition as a result of meeting the needs of another billion or so people. But in this pivotal period when cold war recedes and environmental crisis takes its place there are several hopeful factors. The first is that there is in the developed world a new culture of environmental awareness. The electorates are being greened. They may not be unduly apprehensive about climate change, but they know about the greenhouse effect and how the state of the world and their own localities impinges on their lives.[8] The other factors are the ending of the cold war and the building of a largely successful international security apparatus in the post-war era. The common instinct for survival has shown itself more powerful than the urge to take risks on behalf of national interest. It has proved possible to reach agreements that eventually ended atmospheric pollution caused by nuclear tests, that prohibited environmental modification and kept Antarctica free of nuclear weapons and military establishments. Governments have not been as irrational and totally careless of life as the experience of the Second World War led many to believe. These are factors which provide the basis on which the new environmental security regime can be built, although, sadly, it will probably take more than one catastrophe to shake the world into acceptance of the sacrifices of sovereignty and treasure which will be needed to make it effective. For the post-war generations in the developed nations reared on the statistics of economic and social improvement, it will not be easy to accept that the twenty-first century threatens humanity with hard times and lowered standards of living. Acknowledgement of humanity's common predicament should shape the moral stance, since the goal will be mutual survival.

Notes

1. *New York Times*, 14 and 19 November 1989.

2. This quotation and those following are from *Environment for Peace*, an address by Dr Mostafa Tolba to the Royal Institute for International Affairs, London, March 1990.

3. Timothy O'Riordan, 'The politics of sustainability' in R. Kerry Turner (ed.), *Sustainable Environmental Management: Principles and Practice* (Bellhaven Press, London, 1988).

4. See introduction to David Pearce, Anil Makandya and Edward B. Barbier, *Blueprint for a Green Economy* (Earthscan Publications, London, 1989).

5. From the summary of *Blueprint for a Green Economy*.

6. Speech at the *Better Environment Awards for Industry*, London, 16 March 1990.

7. *Guardian*, 15 March 1990.

8. For a British example, see the *Observer*/Harris poll published in the *Observer*, 15 April 1990. Ninety-two per cent of those polled were aware of the greenhouse effect, but 48 per cent expressed themselves as 'little worried', 22 per cent as not worried at all, and 28 per cent were 'very worried'.

Explanatory Notes

This book is not primarily about climatology or science (although, obviously, both are very much involved) and it does not attempt detailed scientific explanations of the greenhouse effect and the problems of the ozone layer. Many readers will be familiar with both, but I have included the following for those for whom brief easy-to-find summaries will be useful.

The Greenhouse Effect

The greenhouse gases are not (apart from man-made CFCs and halons) newcomers to the atmospheric scene. They are essential atmospheric cladding which prevents the earth from becoming a frozen planet. The current problems are the result of the cladding becoming too dense and warm, largely (it is generally assumed) because of human activities.

The image of cladding, although apt, exaggerates the volume of the gases. They are a very small part of the atmosphere, of which nitrogen and oxygen constitute 99 per cent by volume. Neither has much effect on the earth's heat balance.

The greenhouse gases allow the sun's short-wave radiation to pass through them, but absorb much of the long-wave radiation which the earth throws back towards space (which is what a greenhouse does). Until the beginning of the industrial revolution, there was a rough balance which kept the earth's temperature more or less the same from one year to the next.

The Gases

A 'natural' greenhouse effect raises surface temperature to an average of 15°C. Without it, the world would be 35°C colder than it is.

The two most important greenhouse gases are water vapour, which constitutes less than 4 per cent of the atmosphere's volume, and carbon dioxide.

Water vapour is a key gas since it forms in clouds, and clouds are the most effective barrier to the escape of heat. However, the role of clouds is one of the big conundrums of climatology. They also bounce back, absorb and scatter incoming sunlight to an unknown extent. No one is quite sure whether there will be more or less of them in a warmer global climate.

Carbon dioxide is thought to have contributed 50 per cent of the greenhouse effect in the 1980s. Annual emissions from man-made sources have increased by between three and four times since 1950. Because of its close connection with the fossil fuels (particularly coal) used in electricity generation, controlling its emission globally presents acute problems. Man-made carbon dioxide from burning fossil fuels, about 5.6 billion tonnes a year, should be seen in the context of the natural carbon cycle of 200 billion tonnes exchanged every year between the atmosphere, the land and living things and the oceans. It is not yet clear how constant that cycle is. However, the rise in carbon dioxide does parallel the growth of industry. Its atmospheric concentration was between 270 and 280 parts per million in 1850. The Mauna Loa observatory in Hawaii measured it in 1958 at 315 ppm. By the end of the 1980s it was about 350 ppm. Annual increase in concentrations is 0.5 per cent. A doubling of the 1850s concentration would, at that rate of increase, take place around 2080. However, if increases in the other greenhouse gases are taken into account, the equivalent of that doubling will take place in or about 2030.

The other greenhouse gases are:

Methane It contributes 18 per cent of the current greenhouse effect. Belching cows, paddy fields, melting permafrost, leaking natural gas pipelines – all can contribute to methane as greenhouse gas. Its presence in the atmosphere has more than doubled since pre-industrial (mid-eighteenth century) times. The annual growth in concentration in the 1980s was 0.5 per cent.

CFCs and halons CFCs are the well-known ozone destroying coolant gases in aerosols and refrigerators. Halons are used in fire extinguishers. They contribute 14 per cent of the greenhouse effect and are increasing at the rate of 6 per cent a year. They can last for up to 110 years (some CFCs for 20,000 years) and are the most pernicious greenhouse gases. Molecule for molecule, they are 10,000 times more potent than carbon dioxide and it has been estimated that if production is not checked they will rival carbon dioxide as the main greenhouse gas in the twenty-first century.

Nitrous oxide Six per cent of the current greenhouse effect. From fertilizers, vehicle emissions and the burning of fossil fuels and vegetable matter. Annual rate of increase in the 1980s: 0.25 per cent.

Tropospheric (surface) ozone Short-lived but increasing at about 2 per cent a year; 12 per cent of the current greenhouse effect. Caused by the photochemical reaction of sunlight with carbon monoxide (much of it from cars) and nitrogen oxides, most of them man-made. Its annual increase in the 1980s was 1 per cent.

Temperatures and Sea-Levels

Temperatures The generally accepted estimate is that the world warmed by 0.3–0.6°C during the last 100 years. The IPCC scientific assessment group predicts that without any measures to check greenhouse gas emissions the increase in the mean global temperature will be about 0.3°C per decade – greater than anything seen in the 10,000 years of the interglacial. This would mean the world being 1°C warmer than it is today by 2025 and 3°C warmer by the end of the century. Under other scenarios in which controls of varying severity are introduced the increases would be from 0.1°C to 0.2°C each decade. By comparison, the mean global temperature during the last ice age is estimated to have been 4°C colder than it is today.

Sea-Level Rises   The subject is extremely complicated and the increases to be expected undergo constant revision. The top 2.5 metres of the oceans have been described as being the equivalent in terms of heat capacity of the entire atmosphere. Different currents move at different depths and with great variations in temperature. So it is difficult to estimate thermal expansion as a result of surface warming. It is equally difficult to work out what will happen to the ice sheets at the poles. A simultaneous thinning and shrinking of the sea ice lessens its albedo, or reflectivity, thereby allowing the oceans to absorb more heat from the sun. On the other hand, global warming is likely to produce more precipitation over the icecaps, notably Antarctica's, which would mean that water was being taken out of the oceans and stored as ice. Latest estimates are that sea levels could rise by about 20 cm by 2030 and by 65 cm by the end of the 21st century. There will be significant regional variations.

(*Sources:* Association for the Conservation of Energy; report of Commonwealth group of experts on climate change; paper by Dr F. Kenneth Hare at 1988 Toronto conference on climate change; report of IPCC Scientific Assessment Group; paper by Dr J.T. Houghton of the Meteorological Office; and Dr Pier Vellinga, Director of the National Climate Change Programme, The Hague.)

The Ozone Layer

The ozone layer is in the stratosphere, the upper atmosphere, between 9 and 30 miles above the earth's surface. It is formed when the sun's ultraviolet radiation breaks up the two atoms of oxygen molecules into single atoms. These then combine with other oxygen molecules to form three-atom ozone molecules, resistant to ultraviolet radiation and able to absorb it. Ultraviolet can cause skin cancers and cataracts (and probably other ailments) in humans and it is damaging to plants and marine life. In 'natural' circumstances (before the introduction of CFCs and halons) the ozone layer was kept in balance through natural processes, including destructive gases released at ground level. The production of harmful man-made gases tipped the balance against ozone. Most ozone is generated above the equator, where the sun's radiation is strongest, and then carried by stratospheric winds around the earth towards the poles. The depletion of the ozone layer was first noted over the Antarctic, and, more recently, over the Arctic.

Every year the world's industries have been producing more than one million tonnes of chlorofluorocarbons (CFCs) and halons for use as refrigerator coolants, in the manufacture of foams, aerosol can propellants, solvents and fire extinguishers. CFCs – the main problem – have been with us since the 1930s, a valued invention because they are non-toxic, non-flammable and cheaper to make than any known substitute. Released, they rise very slowly (they can remain in the troposphere, or lower atmosphere, for 100 years) and intact to the stratosphere. Once there, they are broken up by ultraviolet radiation, releasing chlorine atoms which destroy ozone, CFCs are also an important greenhouse gas (see above).

CFCs and Halons

The main CFC gases are:

> *CFC-12*. This is responsible for nearly half the ozone depletion. It is used in refrigeration, air conditioning, aerosols and foams.
> *CFC-11*. Foams, aerosols and refrigeration.
> *CFC-113*. Solvents.

International co-operation on protecting the ozone layer has been good. The Vienna Convention on its protection was negotiated in 1985 and followed in 1987 by the Montreal Protocol, which provides for the control of CFCs and halons. The London conference in June 1990 agreed to tighten up the protocol and phase out production and consumption of CFCs by not later than the year 2000.

(*Source:* Introduction to the ozone layer issue, Saving the Ozone Layer Conference, London, March 1989.)

Select Bibliography

Much of the most absorbing material on climate change and the environment is contained in the reports of parliamentary and congressional hearings and in conference reports, but reading them requires time and dedication. I have limited myself to listing a small number of publications which may be of interest to the reader interested in exploring further into the terrain covered by this book.

Ecology in the 20th Century: a History. By Anna Bramwell. Yale University Press, New Haven and London. 1989.

Climate, History and the Modern World. By H.H. Lamb. Methuen, London. 1982.

Climate and History: studies in past climates and their impact on Man. Ed. by T.M.L. Wigley, M.J. Ingram and G. Farmer. Cambridge University Press. 1981.

International Environmental Policy: Emergence and Dimensions. By Lynton K. Caldwell. Duke University Press, Durham, N. Carolina. 1984.

Climate Change and World Affairs. By Crispin Tickell. Co-published by The Center for International Affairs, Harvard University, and University Press of America. Revised edition, 1986.

The International Politics of Antarctica. By Peter J. Beck. Croom Helm, London. 1986.

Our Common Future. Report of the World Commission on Environment and Development. Oxford University Press. 1987.

The Greenhouse Effect: Negotiating Targets. By Michael Grubb. Royal Institute of International Affairs, London.

Blueprint for a Green Economy. By David Pearce, Anil Makandya and Edward B. Barbier, Earthscan, London. 1989.

The Coming of the Greens. By Jonathan Porritt and David Winters. Fontana, London. 1988.

Green Parties: an International Guide. By Sarah Parkin. Heretic Books, London. 1989.

No Timber Without Trees: Sustainability in the Tropical Forest. By Duncan Poore and others. Earthscan, London. 1989.

Saving the Tropical Forests. By Judith Gradwohl and Russell Greenberg. Earthscan, London. 1988.

Gaia: A New Look at Life on Earth. By J.E. Lovelock. Oxford University Press. 1987.

State of the World, annual reports of the Worldwatch Institute pub. by W.W. Norton, New York and London.

Natural Disasters: Act of God or Acts of Man? By Anders Wijkman and Lloyd Timberlake. Earthscan, London. 1984.

Climate Change: Meeting the Challenge. Report by a Commonwealth Group of Experts. Commonwealth Secretariat, London. 1989.

Turning up the Heat: Our Perilous Future in the Global Greenhouse. By Fred Pearce. Bodley Head, London. 1989.

The Greenhouse Effect: A Practical Guide to the World's Changing Climate. By Stewart Boyle and John Ardill. Hodder and Stoughton. 1989.

Index

acid rain 27, 45, 52, 53–4
acidification, soil 54–5, 187
AEA 223
Agassiz, Louis 38
agriculture 118, 187
    adaptation 120
    Russian 117, 118–19, 120
    set-aside 121
    surpluses 120–1
    US 117–18
aid 205–7
Alaska 24
    oilfields 84, 177
albedo 39, 81
algal bloom 71
Allen, Robert 31
aluminium 53
Amazon Pact 98, 100
Amazonia 95, 97–8
ammonia 54, 186
Antarctic Minerals Bill 85
Antarctic Treaty 79–80
Antartic world park 86–8
Antarctica 43, 75–88
    circumpolar current 61
    climate change 80–1
    icepack 43
    ice-sheet 37, 63, 75
    military use 76
    mineral resources 77, 84–7
    national claims to 75–6
    non-military use 79
    strategic value 76–7
    tourism 79, 88
Aral Sea 122
Arctic 51, 114

oilfield 84
Argentina 80, 85
Arrhenius, Svante 37
atmosphere 43
    modification 51
atmospheric nuclear testing 35
Australia 13, 80, 85, 86
Austria 175

Baltic Sea 71
Bangladesh 21, 64–5, 69, 124
Belgium 80, 86, 175
Bernthal, Fred 46
Bhopal 199
biosphere 28, 43
birth control 28, 68
Bloomfield, Lindsay 137
Blueprint for Survival 31, 32
Bolivia 103
Borja, Rodrigo 98
Boyle, Stewart 163–4
Brady plan 21
Brazil 6, 10, 19, 83, 92, 98–100, 112,
    166, 205–6
Brélaz, Daniel 132
Brezhnev, Leonid 52, 56
Britain 5, 6, 17, 38, 47, 57, 58, 73, 80,
    85, 98, 164
British Antarctic Survey 81
Brower, David 149
Brown, Lester 121
Brundtland Commission 21, 47, 159,
    217, 218
Bryce, Stewart 165
Budyko, Mikhail 121–3
Burke, Tom 145

Bush, George 2, 12, 47, 57–8, 115
Byrd, Richard 78

Callendar, G.S. 37
Canada 52, 54, 57, 108, 123
cancer
    oesophageal 206
    skin 107, 111, 113
    see also leukaemia clusters
carbon dioxide 2, 6, 20, 37, 154–68,
    194, 228
    burning, effect of 91
    emission cuts 6, 9, 17–18, 145, 159,
    160, 161
    man-made 17, 38, 160, 179
carbon monoxide 194
carbon tax 206
cars 192–7; *see also* passenger transport
Carson, Rachel 25, 27, 32
Carter, Jimmy 25, 119
catalytic converter 193–4
cataract, eye 108
Central America 95, 96
Chalker, Lynda 209
Channon, Paul 192
Chernobyl 24, 176
Chidzero, Bernard 209
Chile 80, 85
China 6, 9, 10, 17, 19, 33, 64, 83,
    114–15, 166, 208
chlorine monoxide 110
chlorofluorocarbons (CFCs) 10, 20, 27,
    99, 105, 106–8, 114, 115, 213, 229,
    231
Cites 101
climate change 3, 4, 13, 42–3, 44, 80,
    117–29
climate, grand 43–4
climate modification 121–4
climate stabilization 159
climatology 14–15, 38, 42–6, 48
cloud cover 37, 45, 46
cloud seeding 52
Club of Rome 13, 30, 31, 222
coal 6, 10, 160, 179
    Antarctic 83
cod wars 72
cold car 3, 12
Collier, John 171
Commoner, Barry 27
Comprehensive Environmental
    Protection Convention 86
Cooke, Lindsay 137–8

cosmic radiation 44
cosmos-climate 44
Costa Rica 103
Cousteau, Jacques 86
CRAMRA 84–5
cryosphere 43, 45–6
CSCE 56
cyclones 64, 66
Czechoslovakia 54, 57, 201

Darling, Francis Fraser 31
debt for nature 103
debt, Third World 20, 103, 206, 208,
    216
debt-servicing 20
Deep Ecology Movement 29
deforestation 20, 93, 94, 100
Delors, Jacques 204, 208
deserts 126, 127
desertification 94, 127–8
dimethyl sulphide 83
dioxin 199
diplomacy 5, 59–60
    environmental 38
Dounreay 17
drought 5, 13, 54, 155
Drozdov, O.A. 122
Dubos, René 26, 27
Dunant, Henri 4

Earthwatch 34
East Germany 33, 57, 200–1
Eastern Europe 11, 17, 199–210
EC 103, 108, 114, 134
ECE 56, 57
economic aid 204–7
economic growth 5, 13
eco-politics 131–42
Ecuador 103
EFR 172
Eggar, Timothy 85
Egypt 64, 65, 112, 124, 214
Ehrlich, Paul 27–8, 32
electricity production 159, 163
emigration 127; *see also* refugees
energy 5, 125, 162
    consumption 159, 162, 165–6
    efficiency 68, 164, 179, 180
    renewable 160, 161
    saving 162
    tax 158
environmental degradation 19
environmental disease 202–3

Environmental Economics Centre 18
Environmental Protection Agency 48
environmentalism 3, 6
equilibrium state 29, 30
ethanol 196
Ethiopia 8, 120

Falklands War 76, 80, 84
famine 5, 32
FAO 92, 96, 97
Farley, Joe 109
FBRs 170–4
  opposition to 173
fertilizer use 20, 94, 125–6
Finland 175
firewood 93–4
Flixborough 199
flue gas desulphurization 57
Food Aid Convention 124
Foratom 175
forest
  closed 91
  death 54
  loss of 92
  managed 96
  open 91
  sustainable 91, 93
  tropical 90–104
fossil fuels 20, 37, 125, 159
France 9, 57, 73, 80, 85, 175
Friends of the Earth 97, 146, 149–50,
  163

Gaia 29
Gandhi, Indira 33
Gandhi, Rajiv 168
Gayoom, Maumoon Abdul 64
Global 2000 25
global commons 50
  Antarctica – *see* chapter 6
  atmosphere – *see* chapter 4
  oceans – *see* chapter 5
  tropical rainforest – *see* chapter 7
Goldsmith, Edward 31
Goncalez, Leonidas Pires 98
Gondwanaland 83
Gorbachev, Mikhail 16, 203
Gould, Bryan 140, 141
grain consumption 118
grain trade 119–20
green conditionality 209, 210
Green parties 7, 132–40
Green politics 7, 134–5, 137

green revolution 123
greenhouse effect 17, 39, 277
greenhouse gases 1, 95, 159, 227, 228
  shares 167
greenhouse theory 37
Greenland 15, 41, 63
Greenpeace 86, 146, 147–9
Group of Seven 47, 168
Grubb, Michael 167
Grünen, Die 132, 139, 140
Gulf Stream 61
Guyana 64

Hague Conference 9
Hague Convention 4
Haig, Alexander 119
halons 20, 99, 106, 112, 229
Hansen, James 17
hardwood 96
  tropical 90, 101
Hare, Kenneth 38
Hargrave, John 132
Hawke, Bob 13, 85, 86
HCFCs 107
Head, Ivan 20
Heap, John 87
Helsinki Final Act 56, 59
Heseltine, Michael 203
Hocké, Jean-Pierre 68
Holmes, Andrew 173
Houghton, John 46
Houston summit 18, 47
hydro-electric power 177
hydrogen 196
hydrosphere 43
hydroxyl 194
hurricanes 63, 66

IAEA 176, 182–4
IBAMA 92, 98–9
ice ages 15, 38–40
Iceland 41, 72
ICSU 79
IGBP 81
IGY 79
India 6, 17, 19, 21, 83, 112, 114, 164,
  166, 208
Indonesia 64, 102
Institute of Terrestrial Ecology 67
insulation, thermal 160, 166
Interdepartmental Group on Climate
  Change 155
interglacials 15, 39

Intergovermental Panel on Climate
    Change 1, 2, 4, 45, 46, 155–6, 163,
    180, 223
International Decade for Natural
    Disaster Reduction 66
International Tropical Timber
    Organization 93
International Wheat Council 124
International Year of the Forest 96
Italy 86
ITTA 101
ITTO 102
Iudin, M.I. 122
ivory trade 101
Izrael, Professor 46

Japan 17, 80, 101, 164
Joint Declaration of the European
    Green Parties 135–6

Karin B 71
Kats, Gregory 178–9
Keating, Paul 85
Keepin, Bill 178–9
Kelly, Petra 132, 139
Kenya 21, 112
Ketoff, Andrea 164
Kruschev, Nikita 118

Large Combustion Plants Directive 57
Law of the Atmosphere 109
Law, Phillip 86
Law of the Sea 70, 71–3, 80, 109
LBGAs 119–20
League of Nations 4
lean-burn engines 166, 194, 222
Lebanon 68
Leigh Pemberton, Robin 207
leukaemia clusters 176, 202
Lindz, Richard 47
lithosphere 28, 43
Little Ice Age 40, 41, 44
Lloyd, Ian 156–7
logging 90, 93
Long-range Transboundary Air
    Pollution Convention 55
Lorenz, Edward 23
Lovelock, James 29
Lubers, Ruud 11, 187

Machta, Lester 37
MAD 12
Maddox, John 32

Mahathir, Dr 83
Maldives 64, 70
malnutrition 94, 208
Malthus, Robert 31, 83, 126–7
Manabe, Syukuru 47
Mansfield, William 109
Marshall, William 171, 182
mass extinction 39
Maury, Matthew Fontaine 61
Maya Peace Park 95
McTaggart, David 146, 148
Meana, Carlo Riga de 205
Medieval Warm Epoch 40
Mediterranean 70–1, 123
Melchett, Peter 147, 153
Mello, Fernando Coller de 12, 205
Mendes, Francisco 98
methane 2, 15, 20, 161, 217, 229
methanol 196
Mexico 20, 112
Midgley, Thomas 106
Milankovitch, Milutin 39
Miracle grain 125, 126
MIT 30
Mitterand, François 9, 131
Moi, Arap 113
Molna, Mario 107
Montreal Protocol 107, 112
Mozambique 101
Mulroney, Brian 25, 57
Mumford, Lewis 26

Nahias, Jerome 47
NASA 109
National Oceanic and Atmospheric
    Administration 47
nationalism 104
natural gas 106, 162, 179
Netherlands 9, 11, 175, 185–97
New Zealand 80, 85
NIEO 216
Nigeria 112
nitrogen oxide 53, 55, 58, 194
nitrous oxide 20, 194, 229
Nixon, Richard 52
Non-Proliferation Treaty 182–3, 185,
    223
Noordwijk Conference 9, 17
noosphere 28
North Sea 24, 71
    oil 177
Northern hemisphere 6
Norway 9, 54, 80, 85, 108

nuclear arms 12, 16
nuclear fallout 34–5, 181
nuclear power 5, 24, 160, 170–84
nuclear reprocessing 172
nuclear testing 4, 35, 146, 148
nuclear waste 183
nuclear winter 9

OAU 71
oceanography 61
oceans 61–74
  circulation 62
  currents 43, 82
  warming 46
oil 87
  arctic 51
  reserves 25
  shocks 2, 25, 180
Olympic Games of Pollution 34
Oppenheimer, Michael 14
O'Riordan, Tim 218
ozone 10, 13, 20, 54, 105–16, 230–1
  depletion 58–9, 110–11
    effects of 111–12
  ground-level 194
  tropospheric 229
Ozone Layer Conference 3, 16, 113

Paarlberg, Robert 117
Pachauru, Rajendra 165
Pardo, Arvid 50, 72
Parkin, Sara 132, 138, 140
Parkinson, Cecil 195
Patten, Chris 18, 22, 145, 181, 195,
  197, 219
pcb shipments 71
peace movement 27
Pearce, David 18, 218, 219
Penney, William 171
Peru 68, 72
Philippines 103
plankton 82, 108
Pleydell, Geoffrey 102
plutonium 171, 173–4
Poland 54, 57, 124, 201–2
polar ice cap 62
Pole, Robert 207
polluters, industrial 6
polluter pays principle 6, 19, 158
pollution 6, 16, 217
  atmospheric 4, 58
population growth 7, 14, 21, 28, 128,
  217

Porritt, Jonathan 145, 147
poverty 2, 3, 7, 19, 21, 94, 104, 217
Prohibition of Military or Any Other
  Use of Environmental Techniques
  52
PWRs 171, 177

Rainbow Warrior 86, 148
rainforests 5, 10, 21, 51, 90–104
Reagan, Ronald 16, 25
refrigerants *see* chlorofluorocarbons
refugees, environmental 67–70, 201
Ridley, Nicholas 18
Rocard, Michel 9, 86
Roland, Sherwood 107
Russia *see* USSR

Sagan, Carl 12
Sahel 128
Sarney, President 10, 205
Saunher, Richard 217
Schell, Jonathan 13
Schipper, Lee 164
Scott, Peter 78–9, 150
sea defence 66–7
sea levels 6, 14, 62, 63, 65, 230
seabed rights 72
Sellafield 174
Seveso 199
sewage 70
Shevadnadze, Eduard 203
Sierra Club 148, 149
Sizewell B 177
slash-and-burn 52, 91, 93, 95, 100
smog 194, 199
soil erosion 93
solar activity 44
solar flares 15, 44
solar radiation 38, 43, 44, 45
South Africa 80, 101
sovereignty 12
Spain 57, 175
Stockholm Conference 33–4, 53
storm-surges 63, 69
Strong, Maurice 33
Suess, Edward 28
Sullivan, Francis 102
sulphur dioxide 45, 53, 55
  in Eastern Europe 202
Sulphur Emissions, Protocol on 204
sulphuric acid 53
Sundararaman, Sam 47
sunspots 44, 45

sustainable development 2, 18–19, 24–5, 68, 218, 222
conference on 217
Sweden 54, 108, 175
Switzerland 54, 175

Teilhard de Chardin, Pierre 28
Thailand 64
Thatcher, Margaret 10, 12, 25, 98, 113, 115, 141, 142, 144, 145, 221
thermal expansion 14, 63
Thirty per cent Club 57
Three Mile Island 176
Tickell, Crispin 53, 67–8
Timberlake, Lloyd 69
Tolba, Mostafa 46, 103, 113, 127, 214–15, 216
Töpfer, K. 200
Toulmin, Camilla 128
toxic waste 71, 141
Trail Smelter case 52–3
transport, passenger 165, 191, 197; *see also* cars
public 187, 196–7
Trippier, David 154
Tropical Forest Action Plan 98
Twyford Down 219–21
Tyndall, John 37

U Thant 30
UKAEA 170, 171
ultraviolet radiation 82
UN 4
UN Conference on Environment and Development 5, 20
UN Development Programme 115
UN Environment Programme 10, 34, 46, 68, 76
UN Stockholm Conference on Human Environment 29
UN CLOS 59, 73
UNHCR 68
UNSCEAR 35
uranium 174, 180
USA 2, 5, 6, 9, 17, 20, 27, 38, 51, 57, 66, 73, 80, 164
US-Canadian International Joint Commission 53
US-USSR trade 119–20

Usher, Peter 112
USSR 9, 11, 16, 38, 51, 80, 117, 119, 202
utopianism 26
UV-B 107

Valdez Bay 24, 88
vehicle emissions 27
Vellinga, Pier 63, 67
Venadsky, Vladimir 28
Vienna Convention 105, 109
Vietnam 27, 51, 64
volcanic activity 37, 45

Wakeham, John 158, 177
Waldheim, Kurt 33
Wallich, Henry 30
WANO 181–2
Warren Andrew 164
wealth 3, 17
weather manipulation 4, 51–5
West Germany 6, 54, 73, 175
Western Europe 17
Wigley, Tom 45
Woods Hole Oceanographic Institute 65
World Association of Nuclear Operators 181–2
World Bank 20, 98, 115, 224
World Climate Conference 4
World Environment Day 152
World Food Programme 129
World Meteorological Organization 46, 48, 82, 114
World Ocean Circulation Experiment 62
World Plan for Action 108
World Wide Fund for Nature 102, 146, 150–1
WRCP 82
WRI 96
Wykman, A. 69

Yugoslavia 57

Zaire 92, 215
Zhakarov, Vladimir 114
Zimbabwe 101